CHAPTER
8

Digital Video 62

CHAPTER
9

Ancillary Devices 76

CHAPTER
10

Networking 84

CHAPTER
12

Multimedia ToolBook   109

CHAPTER
13

Developing a Multimedia Program   129

PREFACE

Personal computers have had a role in higher education ever since they became available. There were many Apple II and Commodore 64 labs on campuses across the nation. Until recently, computers have been used as tools, but less often as instructional tools. Students might have used the computer to write a term paper, create a report, or even put together a resume. They might have used it to create a diagram or to assist in the solution of engineering or science problems. This is starting to change. Now personal computers have legitimate roles as instructional tools. There are many factors causing this change. One of the most important of these factors is multimedia.

Multimedia, which includes the use of text, graphics, sound, images, and video, combined with user interaction, has enabled authors to create compelling programs that engage the student's higher-order thought processes, resulting in measurable increases in learning.

The entertainment industry has also discovered the power of multimedia on a personal computer. Retail software stores dedicate a considerable amount of shelf space to computer-based games. This phenomenon has helped foster the use of computers in higher education, as the volume of games sold has helped decrease the cost of computers, advance the state of the art in multimedia technology, and make available in the marketplace sophisticated multimedia authoring tools.

The purpose of this book is to provide a general introduction to multimedia. It is aimed at students who are taking an introductory computer, software, or multimedia course. The aim is to explain the hows and whys of multimedia, while leaving out most of the technical details. The book is also appropriate for faculty who are interested in learning more about multimedia. There is a logical progression to the topics. After an introductory chapter, we discuss why computers are digital and then explain how multimedia material, which is analog, is converted into digital form (Chapter 2). You will learn that this conversion results in large data files, which leads to the topic of compression (Chapter 3). From there, each of the multimedia technologies receives separate coverage. Throughout the text, examples of current software products are used to illustrate the manipulation and use of the multimedia material. Chapter 9 discusses devices such as scanners, tablets, and projectors, which are necessary adjuncts to the use of multimedia. Multimedia networking and the Internet are discussed in Chapter 10. Chapter 11 presents a survey of multimedia tools and software. Chapter 12 is dedicated to a specific product, Multimedia ToolBook. The purpose is not to provide you with ToolBook programming skills, but to give you an idea of how it all comes together in a multimedia program. Chapter 13 offers hints and suggestions for those who wish to undertake a multimedia project.

This book has its origins in a white paper titled "Introduction to Multimedia Featuring the Advanced Academic System," which was created to explain the specific multimedia features of that product. That paper evolved into another paper, "Introduction to Multimedia." There were many requests from academics for a copy of that paper; the decision was made to publish it in the form of this book, to make it more easily available. I hope you feel it was worth it.

ACKNOWLEDGMENTS

I want to thank several people without whom this book would not have been possible: Stanley Smith of the University of Illinois; William Graves and Jim Noblett, both of the University of North Carolina at Chapel Hill, for helping me understand how computers can contribute to the teaching and learning process; Steve Griffin and others at the IAT for technical help; Samuel Abraham of Siena Heights College, Donald L. Jordan of Lamar University, Robert Perkins of the College of Charleston, and David C. Yen of Miami University for their helpful technical reviews of the manuscript; Kathy Shields and Tamara Huggins of Wadsworth for their support of the project; and Gary Mcdonald of Wadsworth and Greg Hubit of Bookworks for taking a draft and transforming it into a readable and attractive finished product.

Two people deserve very special thanks: Rita, my spouse, lover, and friend, for providing suggestions, encouragement, and the gift of time, so precious in a family with young children; and Diana Oblinger, manager and friend, who encouraged the work and made it possible, and who contributed to some of the book's content.

1

Introduction

The past few years have seen an incredible growth in the use of personal computers in higher education. Many factors can be credited with fueling this growth, such as great increases in performance coupled with decreases in cost and the increasing availability of sophisticated authoring tools. One of these factors is undoubtedly the emergence of **multimedia** capability on personal computers. This book will serve as a solid introduction to the concepts underlying multimedia. You will understand what is unique about multimedia, what the challenges are, how to get started with it, and why it works the way it does.

After this initial orientation to multimedia, you will begin to recognize that the range of possible **peripheral devices,** adapter cards, display technologies, and enabling software involved with multimedia is constantly changing. To stay abreast of these technologies, we suggest you read publications such as *NewMedia, PC Week, PC World*, and *InfoWorld*. In addition, the Institute for Academic Technology[1] publishes a number of technical papers, technology primers, and resource information guides covering various topics in academic technology.

What Is Multimedia?

Some say multimedia represents the biggest revolution in education and communications since the invention of the printing press. Whatever you believe, we all understand that computing has altered our ability to manage information. At its best, computing has shortened the distance between people and information (both textual and numerical). Multimedia allows computing to move from text

---

[1]The Institute of Academic Technology is a partnership between IBM and the University of North Carolina at Chapel Hill, dedicated to facilitating the use of technology in teaching and learning. The Institute can be reached at (919) 560-5031.

and data into the realm of graphics, sound, images, and full-motion video; thus, multimedia allows us to use the power of computers in new ways.

Although many definitions can be provided for multimedia, we have come to think of it as having two key elements:

1. Presentation of information through text, graphics, audio, images, animation, and full-motion video; and
2. Nonlinear navigation through applications for access to information on demand.

The second feature of multimedia mentioned above, nonlinear navigation, is often termed **hypermedia.** Hypermedia represents the ability to move through information nonsequentially, which frees professors and students from linear movement through information, such as going from page to page in a textbook. Instead, faculty can respond to spontaneous questions in class, or students can choose a path more suited to their interests and abilities. The additional flexibility of hypermedia encourages students to return to lessons as many times as they need or want.

In general, there are three main characteristics of hypermedia systems:

1. Hypermedia systems allow huge collections of information in a variety of media to be stored in extremely compact forms, as well as accessed quickly and easily. Thus, comprehensive and diverse materials can be assembled and delivered to learners.
2. Hypermedia is an enabling rather than a directive environment, offering unusually high levels of control. Not only does hypermedia offer new ways to present and learn course content, it also introduces opportunities to diverge from a linear path; to juxtapose text, animation, and sound; and to use technology to aid in reviewing, studying, and producing new interpretations of the content.
3. Hypermedia offers the potential to alter the roles of teachers and learners, and to enhance the benefits and frequency of their interaction. Hypermedia encourages students to become scholars, that is, more in control of their own learning. Teachers can define their role more as coaches than as lecturers.

The term *hypertext* has also been used. **Hypertext** consists of the capability of navigating in a nonlinear way through text. Hypermedia is a more comprehensive term, covering both hypertext and nonlinear navigation through multimedia.

Let's briefly look at each of the multimedia technologies currently available in personal computers and used in higher education:

Images

Pictures, drawings, illustrations, and so on to help explain or exemplify the text.

CD-ROM Audio

A compact disc generates audio using a CD-ROM drive attached to and controlled by a computer.

Digital Audio

The computer generates audio by interpreting specially coded data files that describe the waveform of the audio signal.

MIDI (Musical Instrument Digital Interface)

The computer generates audio by interpreting specially coded data files that describe a sequence of notes and musical instrument selection.

Animation

Computer graphic images are played back in a sequence.

Analog Video

The computer plays back video from an analog source, such as a videodisc player or a video cassette recorder. Audio tracks from the playback device are normally routed to external amplifiers and speakers.

Digital Video

The computer plays back video stored in digital format on a file on a computer disk, a CD-ROM, or a network. Audio tracks are stored in the same file and are played through the computer's audio circuitry in the same manner as digital audio.

Organization of This Book

This book would be incomplete if it only discussed the various multimedia technologies on the market. In order to really understand multimedia, it is important to understand why multimedia data is different from traditional computer data (a word-processing document, a spreadsheet, or a graphics file). Thus we will start out by taking a look at how multimedia data is generated. This will make it clear why multimedia data is unique, leading us into the next topic, data compression. Data compression is necessary in order for today's computers to deal with the massive amounts of data that multimedia entails.

After these general topics, we will examine in more detail each of the technologies mentioned above. We will discuss the specifics of the technology, as well as some current academic applications that successfully employ it. We will see examples of products that implement the technology.

The size of multimedia files introduces yet another problem for the campus. It is not feasible to duplicate the data on many computers (say, each computer in a lab); how can such massive amounts of data be transmitted across a network in a timely fashion without bringing the network to its knees? We will examine some of the ways to accomplish this. Finally, in Chapter 13 we will look at how to get started in multimedia.

You might be a faculty member thinking about writing some courseware that incorporates multimedia, or a student learning about multimedia. Or you might be an administrator in charge of planning a new lab or a network and want to make sure that it can handle multimedia. This book will help you. This book approaches multimedia from a Microsoft Windows point of view, although the principles are the same for multimedia on other platforms.

Why Multimedia Is Important in Higher Education

Our traditional interaction with computer-based information has not been "natural"; that is, it has primarily been through text. In contrast, most of our daily experiences are through sound and images. Multimedia can be thought of as using a computer to provide a multisensory experience. This "experience" is extremely useful in a lecture presentation, as part of a laboratory, or as individualized instruction where the multimedia is controlled and managed by the participant's actions or decisions.

Much of our education has been based on words and numbers. Although we are used to dealing with a text-based world, it is not "natural" in the sense that text not only limits the scope of information we can grasp but also can make some topics more difficult for us to understand because text requires the brain to continuously code and decode information. This limits the speed and range of our communication. Multimedia can help by bringing together sights, sounds, text, and images in a single communication medium. Think, for example, of a biology textbook. No matter how colorful, a book is a poor substitute for the motions and sounds that abound in nature.

The hypermedia aspect of multimedia is also important because it allows students to explore topics at their own pace and in their own manner. Thus each student can tailor the learning experience to his or her own learning style.

Research indicates that multimedia can aid students. At the Marine Biological Laboratory at Woods Hole, Massachusetts, Dr. Daniel Alkon and colleagues showed that the memory process is aided when the brain receives multiple related stimuli over a short period of time. Thus a picture is more effective than text because it is a more natural experience than text. Adding motion and sound involves the addition of the sense of hearing and of multiple stimuli over time. Adding interactivity involves the higher centers of the brain, further enhancing the learning process.

We retrieve memories by matching present input (say, a question on a test) to stored patterns. The more complex the pattern, the greater the number of "keys" with which the information can be retrieved. This is why we can more easily remember information that is associated with other information. Essentially, receiving a stimulus that matches part of the pattern causes our memory to retrieve the complete pattern, in other words, the remembered information. Multimedia, by providing more information to the student, can aid in forming more complex memory patterns in the brain, and thus contributes to the ease with which these memories are retrieved.

No technology, including multimedia, is immune from abuse. Bad examples of the use of multimedia are not hard to find. In many cases multimedia is added to an application just so the author can claim credit for an additional feature. On the other hand, it is also easy to find many applications in which multimedia contributes to the learning experience.

Let me entice you with a few such examples:

- The Microsoft Bookshelf product includes a dictionary. Clicking on a word in the dictionary causes that word to be spoken using the audio capabilities of the computer. Thus students are able to learn to pronounce the word correctly using a technique (hearing the word) that is easier than trying to use a pronunciation guide.
- The Perseus Ancient Greece program (developed by Dr. Gregory Crane at Harvard University) includes a wealth of high-quality images of ancient Greek vases, coins, and other art objects. Some vases are represented by several dozen pictures, showing different views of the vase, as well as many details. The objects represented in Perseus are scattered in museums in different parts of the world, including the United States, the United Kingdom, Germany, and Greece. Thus, in one program students can study objects from ancient Greece that they could only see otherwise by extensive travel. Furthermore, students might not be able to see as much detail in the real

vase as they can see in the photographs because they will probably not be able to get that close to the vase, turn it around, and so on. A student's study of Greek culture, language, or literature is thus enhanced by the availability, in a convenient form (CD-ROM), of a wealth of enrichment materials.

- The chemistry programs authored by Dr. Stanley Smith and Dr. Loretta Jones employ motion video to illustrate chemical reactions. In this way students can study a variety of reactions that are too dangerous or expensive to perform in a wet lab. The programs are flexible enough to allow students to select their own reagents, concentrations, and amounts, so although every student is using the same program and identical videodiscs, they each obtain individual results.

- Perhaps the most compelling example of the effectiveness of the multimedia experience is the use of flight simulators for pilot training by the airlines. The flight simulators provide a realistic experience, including sound, visuals (instrument readings and out-of-window dynamic views), tactual feedback (the feel of the controls), as well as other sensory information (aircraft attitude, acceleration, vibration, etc.). Simulators are cost-effective when compared to training on the aircraft itself, and they are safer learning environments. In addition, simulators can expose pilots to training situations that would be difficult or unsafe to provide with the use of the actual airplanes (e.g., the simulation of mechanical failures).

2

Introduction to Digital Multimedia

We live in an analog world. Despite what the thermometer on a bank's sign might say, temperature does not change in one-degree increments. If we had a precise thermometer, we could track temperature change in tenths of a degree. At considerable expense we might purchase a thermometer that registered hundredths of a degree, or even thousandths, and we could track the temperature to that degree of precision. No matter how fine our scale, we are still not able to capture all of the values a property (such as temperature) goes through. A car accelerates smoothly from zero to fifty-five miles per hour, going through a perhaps infinite series of velocities in between. This analog behavior (a smooth change of values for an object's property) is exhibited everywhere in nature. This smooth behavior does break down at the atomic level, but that is far too small a scale for our senses or most of our instruments to detect.

Why Digital?

Yet the computers we employ are digital, not analog. Why do we choose to represent analog phenomena on digital machines? The reason is that it is cheaper and much more feasible to maintain the accuracy of digital data than of analog data as the data gets processed and transmitted. Suppose we have an analog representation of a musical passage. This representation is in the form of a waveform that tracks the vibration of, say, an air molecule (or a human eardrum) exposed to the sound of the musical passage. If we are now to transmit this signal over a wire, either within the computer or one that connects two computers hundreds of miles apart, the signal is susceptible to distortion caused by electromagnetic noise in the environment, quantum effects, reflections, and so on. If we design the circuits very carefully, such distortion might not be perceptible to the ear. However, after many successive transmissions, the signal would deteriorate to the point that it would no longer be recognizable. It is also very difficult to look at the received signal and determine what is distortion and what faithfully represents the original signal.

In a digital computer all signals and data are represented by one of two states, 0 and 1. Assume that we define that an electrical level of 0 volts is represented by a 0, while a level of 5 volts is represented by a 1. If we transmit the signals over a wire, they will be distorted just as their analog counterparts were. Suppose the 0 arrives at its destination with a value of 1 volt. Since 1 volt is closer to 0 than it is to 5, the circuit has no trouble determining that it really represents a 0, and will pass a 0 on to the next circuit. Likewise a 1 (5 volts) might arrive at its destination with a level of 4 volts. Again, since 4 volts is closer to 5 than to 0, the receiving circuit has no trouble interpreting the signal. Even if the signals, originally at 0 and 5 volts, were distorted to such an extent that they arrived as 2 and 3 volts, the receiving circuit could still correctly reconstruct the original signal. This is illustrated in Figure 1. Assume that a perfect digital circuit is instructed to generate the binary sequence 10011101. The signal would look like the first graph in the figure. The tick marks on the horizontal (time) axis represent the center of the clock cycle, meaning the point in time at which that particular bit of the sequence is guaranteed to be valid. The lower graph represents the same signal, now severely distorted. At the center of each cycle (the tick marks on the horizontal axis) we look at the value of the signal (the square symbols on the graph). If it is above 2.5 volts (the dotted horizontal line) we will consider it to be a 1 and if below 2.5 volts, a 0. Notice that although the received signal looks quite a bit different than the original signal, we can still correctly determine its value.

No matter what the signal is on the wire, it can only represent one of two states, as opposed to one of an almost infinite number of states, as is the case with an analog signal. Therefore, despite the imperfections of our circuits, and the inevitable distortion caused by nature, it is still quite easy to reconstruct the original signal. This fact might be of great comfort the next time you are riding on an airplane feeling its way through clouds and fog, when the pilot must put all his trust in his digital computers and instruments.

FIGURE 1   *Distortion of a Digital Signal*

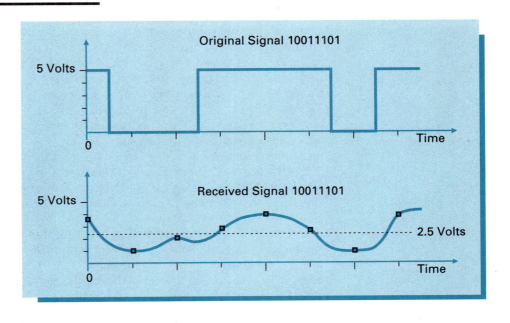

Another advantage of the digital format is that multimedia data in digital form is much easier to manipulate than the same data in analog form. Such manipulation might consist of adding reverberation to a sound file, adjusting the colors and contrast of an image file, or splicing two movie segments from different files together with a transition effect between them. Because the data is in the form of numbers, well-understood **algorithms** can create the desired effects by transforming the numbers, something computers are very good at doing.

ERROR CORRECTION AND DETECTION

Another aspect of digital circuits is that it is easy to add redundancy to the signal so that the receiver can determine if the signal changed in transit. The simplest of these schemes is **parity,** where an extra **bit** specifies whether the signal has an odd or even number of 1 bits. The receiving circuit counts the number of 1 bits received and checks the parity bit to see if there is agreement. If not, an error has occurred. Parity check circuits are used inside many PCs to ensure the integrity of data traveling from the **memory** to the **processor.** Parity is also used in **serial communications.** (e.g., connections of computers over phone lines). Much more elaborate schemes exist and are used not only to determine that an error has occurred, but to correct it on the fly. Such schemes are used in **mainframe** computer memories, **disk drives,** and compact discs.

SAMPLING

Given the fact that we use digital computers, how do we convert analog signals, such as a sound wave, into a digital representation suitable for the computer? It is done by a process called **sampling.** We will look at how sampling is done for an audio signal. The concepts are the same for capturing other types of multimedia, such as photographs and movies.

We all have an idea that sound is caused by vibrations in the air, which in turn cause our eardrums to vibrate. These vibrations are converted into nerve impulses in the inner ear by a process that is still not well understood. Imagine, now, a strip of paper that moves at a constant velocity perpendicular to our eardrum. Imagine also a tiny pen attached to our eardrum. As the eardrum vibrates, the pen traces its movement on the moving strip of paper. If we examine the strip of paper, it might look like Figure 2. The horizontal axis represents time, while the vertical axis represents the amplitude of the vibrations of the eardrum in response to sound.

This is the analog representation of the sound. The signal changes smoothly, as the eardrum cannot change positions instantly. The distance between peaks in the signal is called the **period,** which is related to the frequency of the sound—shorter distances mean higher frequencies, more cycles fitting into a unit of time. The distance of the signal from the horizontal axis represents its **amplitude**—how loud the sound is. The waveform pictured in Figure 2 is a very simple waveform. Waveforms from natural sounds, be they the rustle of leaves in the breeze, human speech, the sound of a single violin, or the sound of a whole symphony orchestra, are much more complex. However, the point is that any sound can be represented by a waveform. Recall that the waveform is simply the trace of the vibration of an eardrum, or an air molecule, or the movement of a loudspeaker's cone.

In order to understand sampling, let's first look at the process graphically. Sampling consists of looking at the waveform at fixed intervals (as measured on the horizontal axis) and noting the value of the waveform (the distance of that point of the waveform from the horizontal axis) at that particular location. Figure 3 shows this process graphically. At specific intervals, the value of the

FIGURE 2 | *Tracing of Movement of Eardrum*

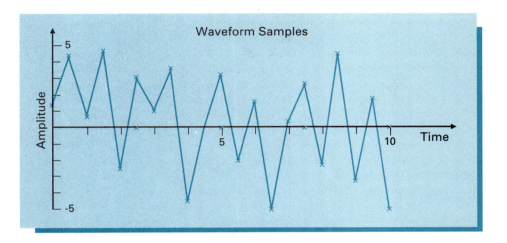

FIGURE 3 | *Sampling the Waveform*

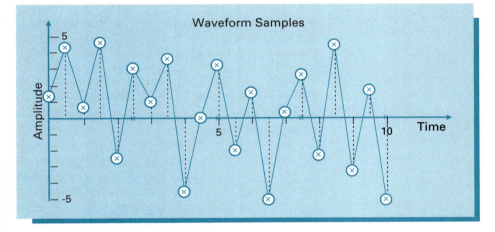

TABLE 1

Digitized Waveform

Time	Value
0.0	1.375
0.5	4.375
1.0	0.750
1.5	4.750
2.0	-2.500
2.5	3.000
3.0	1.125
3.5	3.500
4.0	-4.500
4.5	0.000
5.0	3.250
5.5	-2.000
6.0	1.625
6.5	-5.000
7.0	0.375
7.5	2.625
8.0	-2.375
8.5	4.500
9.0	-3.250
9.5	1.750
10.0	-5.000

waveform at that point (the height of the vertical line connecting the data point to the axis) is noted (the circles represent the sample points).

That value, a number, is then written down. When we finish sampling the waveform we have a list of numbers (Table 1). Playing back the recorded sound is analogous to a game of connecting the dots. The list of numbers is read back. Using the same scale as used for the sampling, a dot is plotted on the graph, its height above the horizontal axis corresponding to the value of the number read. Once we have plotted all the points (Figure 4), we simply connect them together to regenerate the original waveform.

The process in the computer is similar. A microphone generates an electrical signal corresponding to the air vibrations. Inside the computer a circuit, called an **analog-to-digital converter (ADC),** continuously converts the analog voltage value of the electrical signal to its digital representation. At fixed intervals of time, the computer reads (samples) the number and stores it in a file. To play back the sound, the computer reads the list of numbers from the file and sends

FIGURE 4   *Recreating the Original Waveform*

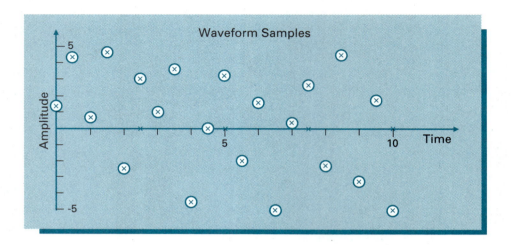

them, at the same rate they were generated, to an electrical circuit called a **digital-to-analog converter (DAC),** which generates a voltage corresponding to the number. This electrical signal is connected to a loudspeaker, whose cone moves in response to the voltage on the line, re-creating the original sound. The technique of sampling and representing an analog signal in this way is called **pulse code modulation,** or **PCM.**

FACTORS AFFECTING QUALITY

There are two factors that affect how faithfully the reproduced sound mirrors the original: sample rate and sample size.

Sample rate is easy to understand. If we do not sample the waveform often enough, we will have fewer dots, which means that when we connect the dots to reconstruct the waveform, our waveform is going to be a poor replica of the original. This is illustrated in Figure 5, where we have one half as many samples as we did in the example of Figure 3.

The question is, how often should we sample? The answer is provided by the **Nyquist theorem,** developed by Bell Labs scientist Dr. Harry Nyquist, which states that in order to preserve all of the information in the waveform, we need to sample it at least twice as often as the highest frequency present in the waveform. Music for compact discs is digitized at 44.1 **KHz,** which gives compact disc audio a frequency response range of 20 KHz, beyond the upper frequency of sound that humans can perceive. In digital multimedia, sampling rates are determined from a compromise between the size of the sound file and the quality one wishes to obtain.

The second factor is sample size. To understand this, let's go back to our example of reading points off the graph (Figure 3). Suppose that for each sample we could only write down its value using a single digit. We would then divide the vertical axis into 10 intervals, and assign each sample a value corresponding to the closest tick mark. This would entail considerable rounding of most samples, so the reconstituted waveform would not closely resemble the original (Figure 6). The error caused by this rounding is called *quantization error*, and its effect is audible in the form of hiss in the resultant sound. If, on the other hand, we could use two digits to store each sample value, then we

FIGURE 5 ▌ *Recreated Waveform with Fewer Samples*

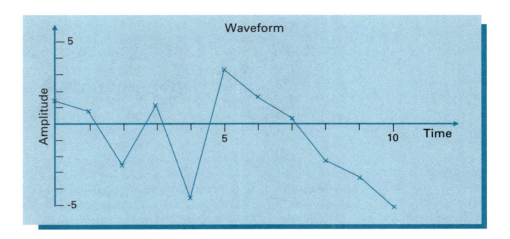

FIGURE 6 ▌ *Recreated Waveform with Smaller Sample Size*

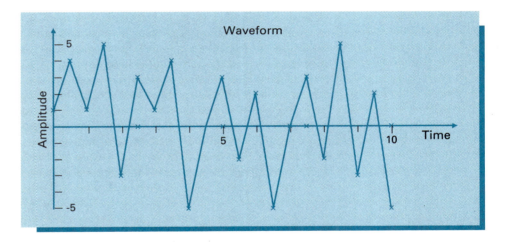

could divide our vertical axis into 100 intervals, and the error per sample would be much smaller. If we used three digits, then we could divide the axis into 1,000 intervals, and the error would probably be negligible, which means that the reconstituted waveform is nearly indistinguishable from the original. However, the larger the number of bits used for the sample, the larger the resultant data file will be. Sample size is an important factor in digital multimedia. Digitized images, for example, can be classified as to how many colors they contain (256; 65,536; 16.7 million; etc.). The number of colors an image contains is a function of how many bits are used to store each **pixel** of the image, in other words, the sample size. For the image examples above, sample sizes are 8 bits, 16 bits, and 24 bits, respectively. When you hear discussions about 24-bit color, it is the sample size they are talking about. Audio is digitized using a number of different sample sizes.

FIGURE 7 | *Aliasing Effect on Wheel*

ALIASING

Aliasing is a phenomenon that occurs when the sampling rate is not high enough to digitize all the information in the waveform; in other words, the waveform contains frequencies that are higher than twice the sampling rate. We are most familiar with aliasing as the stair-step shape of diagonal lines on low-resolution computer screens. Essentially, there are not enough samples (in this case, pixels on the screen) to accurately represent the line. The effect of aliasing is to introduce, in the reconstructed waveform, signals that were not present in the original.

One does not have to look at a computer to see the effects of aliasing. The next time you see an automobile ad on television, or watch a Western movie (one with a wagon in it), look at the wheels. In some cases, although the vehicle is moving forward, the wheel seems to rotate backwards. Wheels with spokes in them facilitate seeing the effect. The wheels seem to rotate backwards because they are rotating too fast with respect to the sampling rate (30 frames per second for television, 24 frames per second for motion pictures). This is illustrated in Figure 7, which shows four frames, side by side, of a movie of a rotating wheel. Between each frame, from left to right, the wheel has rotated through an angle of 345°. Yet we perceive it as rotating backwards by 15°, because the brain is fooled by the proximity of the lettering on the wheel between adjacent frames. The reconstituted signal (the movie or television picture) has extraneous information in it (the look of the wheels turning backwards) that was not present in the original. If you think about this example you will probably achieve a qualitative understanding

FIGURE 8 | *Aliased Signal Caused by Under-Sampling*

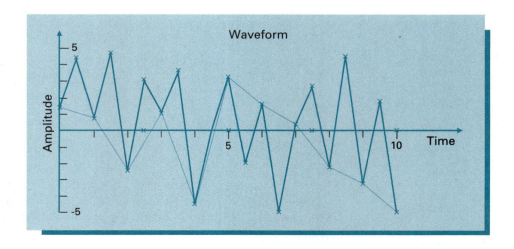

of the Nyquist criterion. If the sample rate is at least twice the frequency, in other words, if there are at least two frames for each revolution of the wheel, the lettering on the wheel will always have moved less than half a revolution between frames, and the brain will correctly interpret the movement of the wheel.

Let us also look at an example with our sound signal (Figure 8). The original waveform is shown as a dotted line, with the original sample points represented as circles. If we reduce our sample rate to one tenth of the original, the resultant signal is a low-frequency signal (solid line) that bears no resemblance to the original.

In order to eliminate aliasing, which would be heard in the reconstituted sound, the signal to be sampled is filtered to remove any frequency components that are greater than twice the sampling rate.

3

Compression

We saw in the last chapter how analog data is converted to digital. What we have not discussed are the characteristics of these files. For the most part, multimedia data files are quite large. Let's consider a few examples:

- Medium-quality monaural sound (22 KHz, 8-bit samples) takes 22,050 **bytes** per second of sound. This means that a minute of sound takes 1,323,000 bytes, which barely fits on a high-density diskette. An average popular song takes about 5 megabytes of storage.
- CD-quality stereo audio takes 176,400 bytes per second, or 10,584,000 bytes per minute.
- A high-quality 640 × 480 pixel photograph takes almost a **megabyte** of storage.
- Full-motion video (30-frame-per-second **NTSC** video) takes about 30 megabytes per second of video. A 20-second video clip would take 600 megabytes, or much more room than is available on the majority of hard disks shipped on personal computers today.

Obviously these large file sizes are a problem for personal computers. These files strain the capacity of the disk drives and the processor's power to move and process the data. The answer to this problem is to reduce file sizes through data **compression.**

Fundamentals of Compression

In some instances we can reduce the size of multimedia files by judicious choices on how the data is captured. For example, if your goal is to record the pronunciation of foreign words and phrases, there is no gain in recording it as CD-quality stereo sound. Monaural sound with low sample rates and small sample sizes will do just fine, making the sound files much smaller and more manageable.

However, this is not really what is known as compression. Initially it might be hard to accept that data can be compressed without something being lost. In fact, many compression schemes are **lossy** methods, meaning that when the data is decompressed it is not an exact replica of the original, but it is probably close enough for our purposes. Other compression methods are **lossless,** meaning that the decompressed data is an exact copy of the original. For example, several programs, such as Stacker, compress the information on the hard disk to give the user more space. When the computer tries to access any of the compressed files, these products intercept the data and decompress it on the fly. By necessity these products use **lossless compression** schemes.

LOSSLESS COMPRESSION

Let's discuss some lossless compression schemes that are used in multimedia and computing.

Run-Length Encoding To help us understand lossless compression, let us consider scanning and compressing the information on a page. Because the information is in black and white, we only need one bit to store each dot on the paper. If the dot is black, the bit will have a 1 value; if it is white, a value of 0. How many bits it takes to store the contents of the page in digital form depends on the resolution of our scanner. Some scan 300 dots per inch, others more, some less. The scanner scans the page serially, just as we read it, a line at a time, assigning a 1 or 0 value to each dot in the line. As we look at the scanned data, we notice there are many 0's, corresponding to all the white space on the paper—the margins, the space between paragraphs, the space between lines, and so on. If our scanner scans at 300 dots per inch, the white space between paragraphs will be represented by thousands of consecutive 0's. What if, instead of blindly storing strings of 1's and 0's, we defined a code that represented a 1, another code for a 0, and followed the code with a number representing how many 1's or 0's follow in the scanned image. Thus we can reduce our strings of thousands of 0's to just a few bits each. For example, a program trying to recreate the image to display it on the screen need only look for the codes and then create as many white or black dots as the number following each code.

If there are places in the image where there are very short sequences of dots of the same color, this scheme will actually result in more data than if we just recorded the dot values themselves. However, on average, the scheme will save us a considerable amount of space. The amount of compression we can achieve depends on the data we are trying to compress. Because all compression schemes share this attribute, one normally characterizes a compression algorithm by its *average compression ratio* or its *compression range.*

Another truism for compression schemes is that there is a compromise involved. We do get smaller file sizes, but in return, some processing is required to compress the data at one end and to decompress it at the other.

The simple, widely used compression scheme just described is known as **run-length encoding,** or **RLE.**

Huffman Encoding Suppose we are trying to compress a text file. In such a file, each character is represented by a number and each number is one byte (8 bits) long. Personal computers have standardized on the **ASCII** code, which represents letters in just such a way. We notice, though, that some letters occur very frequently, such as *t, r, s, a, e,* and so on. Others, such as *x* and *z,* occur very infrequently. What if we were to develop a compression algorithm where the most frequently used letters were represented by short bit strings (2 to 4 bits) while the least used characters were represented by longer bit strings? Although it would take more bits to represent these infrequently occurring letters, that increase is more than made up by the fact that we save bits on the frequently occurring letters. Overall, the size of the file is reduced. This technique is known as **Huffman encoding,** and it too is widely used. Fax machines use a combination of RLE and Huffman to achieve significant compression ratios. The processing involved is inexpensive compared with the cost of the telephone call, so it is a good trade-off in order to send the least amount of data possible, resulting in the shortest connection time.

Other Simple Lossless Schemes   Another approach, used in some disk compression products, involves identifying common patterns or strings in the information, and finding places where that string or pattern occurs again. An example of a pattern might be words in a text file. Each word is stored only once. Every other time it occurs, a pointer is used to point to the word. When the decompression software is reading the file, and it comes upon a pointer, it simply looks up the word (or string or pattern) the pointer is pointing to, and copies that in place of the pointer in the decompressed file.

This type of compression brings up a general point: The fact that we can compress files is not a fortuitous occurrence but a result of how information is represented. The more rules we place on information, the more we restrict the number of different forms it can take, which means we can compress it more. For example, we could compress a text file on a letter-by-letter basis. However, we would obtain better compression if we compressed on a word basis, because not all combinations of letters are valid words (this fact has the effect of placing rules on the representation of words). Taking this example further, we could compress sentence-by-sentence, as not all sequences of words are valid sentences. We would achieve higher compression, but the task of assigning codes to every possible sentence would be overwhelming.

LOSSY COMPRESSION

In **lossy compression** we recognize that a loss of information, reflected in the fact that the decompressed picture, movie, or sound will not quite match the original, is acceptable. We might, for example, compress a picture by discarding every other pixel. The resultant picture will not be as good, but it probably will be quite acceptable.

Another approach is to realize that some properties in a multimedia object are more important than others. We perceive luminance (brightness) better than color, so perhaps we can sacrifice some color information and preserve more brightness information.

In the end, it actually does not matter whether or not the reconstituted object is an exact replica of the original. What does matter is how close we perceive them to be. For example, when we hear a loud sound at a certain frequency, our ears are deaf, for a period of time, to other sounds close in frequency to the loud sound. Therefore, a compression algorithm could discard such sounds following the loud sound for the specified period of time, thus saving space. Although the original and the reconstituted files would not be the same, the user would not perceive them as different.

JPEG   The Joint Photographic Experts Group, or **JPEG,** is a standards body that defined a very comprehensive specification for the compression and representation of image (photographic) data. The simple schemes we discussed above don't work very well on photographic images; thus the need for a different scheme. JPEG uses an algorithm based on some complex mathematics called a *discrete cosine transform.* Let's discuss, without getting into the technical details, how this algorithm compresses data. Take, for example an 8 × 8 block of pixels in a picture. This block is composed of 64 pixels, and each pixel is represented by a number. The first number corresponds to the first pixel, the second number to the second pixel, and so on. The discrete cosine transform generates a different set of 64 numbers (coefficients) based on the 64 initial pixel values. These new numbers don't represent the color values of each individual pixel. Instead, they represent frequency components of the image. The first number corresponds to the lowest-frequency component, the second number to the next lowest-frequency component, and the last number to the highest-frequency component. There is little high-frequency content in most photographs, so the high-frequency coefficients can be discarded with little loss of picture information. After transformation, the coefficients undergo a process called *quantization*, in which the number of bits used to represent each coefficient is reduced by assigning a single value to similar numbers, much in the way that we round decimal numbers, for example, from 2.51 to 2.99 up to the integer 3, and the numbers 2.01 to 2.49 down to the integer 2, using only one digit to represent the number instead of three digits. The JPEG algorithm can compress image files by a factor of 10 to 20, with little perceived loss in image quality. It also allows us to specify compression parameters, so we can trade off loss of quality for compressed file size. An image may be compressed by a factor of 100; although the loss in quality will be apparent, it might still be acceptable. The drawback is that the JPEG algorithms are very compute-intensive, meaning that they take a significant amount of processing.

The following set of figures (Figures 9–11) illustrate the effects of JPEG compression on image quality. Table 2 compares the file sizes for the three images. Figures 12–14 show blowups of a portion of the images in Figures 9–11, allowing you to see the effects of compression in greater detail. Notice the blocky appearance of the compressed images, a characteristic effect of JPEG.

FIGURE 9   *Original Image*

FIGURE 10   ❘   *Image of Figure 9 with 58% JPEG Compression*

FIGURE 11   ❘   *Image of Figure 9 with 95% Compression*

Because of its status as an international standard, JPEG is gaining wide acceptance. It is supported in Apple's QuickTime and in a number of hardware and software products, such as CorelDRAW!, Asymetrix ToolBook, and others. Furthermore, some products use JPEG as the form for storing their own pictures. The Microsoft Scenes collections of screen savers are an example.

TABLE 2   *Comparison of Image File Sizes*

Figure	Image	File Size (bytes)
9	Uncompressed	99,382
10	JPEG Compressed	42,243
11	JPEG Compressed	5,088

Movies   Movies (motion video) represent an enormous challenge for the personal computer. As previously mentioned (page 14), motion video requires some 30 megabytes of data per second. Motion video is composed of a sequence of still images (called **frames**) that play at the rate of 30 per second, along with a sound track that can have multiple channels of audio.

The fact that motion video is composed of a sequence of images opens the way for another form of compression, called **interframe compression.** In a sequence of images, it is often the case that not much changes from frame to frame, except when the scene shifts. For example, during an interview or news show the face of the person talking will move, but the background will probably not change. If the video is following a moving object, such as a car, the background is changing but the image of the car within the frame will vary little from frame to frame. In interframe compression the algorithm only saves the information that changes from frame to frame, not the whole frame. However, since each frame is compressed using lossy compression, after a number of frames the errors accumulate enough that quality is no longer acceptable. Thus, at specified intervals, a complete frame (one that does not depend on the previous frames) is used. This frame is called the *reference frame*. Thus two types of compression take place for movies: **Intraframe compression,** where the information in a

FIGURE 12   *Detail of Image of Figure 9*

FIGURE 13   *Detail of Image of Figure 10*

FIGURE 14   *Detail of Image of Figure 11*

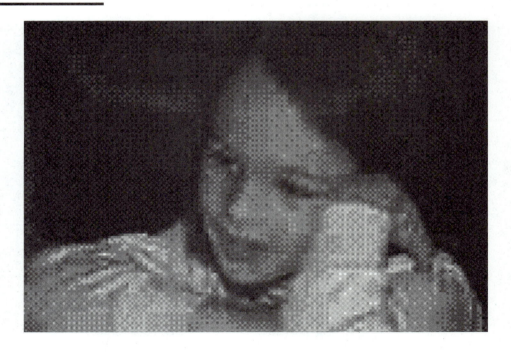

frame is compressed as if it were a still picture, and interframe compression, which is the compression from frame to frame. Both types of compression are quite elaborate, and the technology is still in its infancy.

A number of compression algorithms exist for motion video. Some rely on the computer's processor to decompress the file when it is being played back; this type of video is known as *software motion video*, because the decompression is handled in software. Examples include the supplied decompressors with Apple's QuickTime, Microsoft's Video for Windows, Intel's Indeo, and SuperMac's Cinepak. Software motion video has the advantage that no special hardware is required in order to play back these movies. The drawback is that the task of decompressing the video and pumping the data through the system at the required rates is just about beyond the capabilities of today's personal computers. Software motion video compromises by playing the video in small windows (either 160 × 120 pixels or 320 × 240 pixels) instead of full screen (640 × 480 pixels) and playing at 10 to 15 frames per second instead of at the full-motion video rate of 30 frames per second. Both of these techniques cut down the data rate significantly, but many users are not satisfied with the results.

Other motion video relies on a specialized hardware card in the computer to assist with playing back the video. The computer's processor handles the reading of the movie file. The hardware card, using a special processor, decompresses the file and mixes the video with the computer's own screen image. These cards enable the playback of 30-frame-per-second (full-motion), full-screen video. The disadvantage of this approach is that this specialized hardware must be installed on every machine on which you want to play the video files. Using these cards, compression ratios of 200 to 1 are achievable.

MPEG   The Motion Picture Experts Group (**MPEG**) is a standards body that is assembling a standard specification for motion video compression. It is a newer standard than JPEG, and uses JPEG techniques for intraframe compression. A number of products on the marketplace use MPEG technology for video compression. Sigma Designs' Reel Magic card uses the MPEG standard to play full-motion video. In time, MPEG might become a widespread standard for motion video compression, replacing the variety of products existing today. MPEG is the leader in hardware motion video standards. There are also software MPEG decompressors, although they cannot sustain full-motion nor full-screen video. Given the complexity of the MPEG algorithms and the need to decompress in real time, most MPEG implementations in the near future will require a hardware card to aid in this task.

AUDIO COMPRESSION

Audio is difficult to compress. It does not lend itself well to simple schemes such as RLE and Huffman, and it is not productive to use complex schemes such as discrete cosine transform. However, a couple of compression methods do apply to audio.

Delta Modulation   Recall our discussion of pulse code modulation (page 10). Our sample size had to be big enough so that we could, with a single number, represent fairly accurately the amplitude of any point in the signal. What if, instead of each sample representing the absolute value of the signal, it represented the change in the signal since the last sample? Furthermore, what if we reduce the sample size to 1 bit, so that it is 1 if the amplitude of the signal increased since the last sample, and 0 if it decreased? This technique of digitizing sound is known as **delta modulation.** Its advantage is that it reduces the sample size from 8 or 16 bits down to 1 bit, thus saving a significant amount of space.

FIGURE 15 | *Delta Modulation*

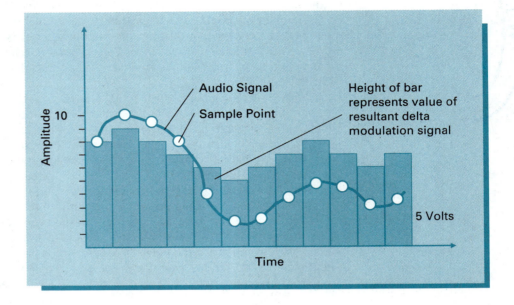

Figure 15 shows an example. At each sample point, the signal (line) is sampled and the output value is either a 0 or a 1, according to the algorithm described above. The vertical bars represent the resultant signal values at each sample point. Note that the difference in height between any two adjacent columns is always a single unit (as measured on the vertical axis), since the algorithm only records an increase or decrease in signal amplitude—there is no recording of the amount of change.

There are several disadvantages to delta modulation. First, because each sample is relative to the last one, any error is going to be propagated throughout, distorting the sound. Second, there is the danger that the signal may change rapidly at some point; delta modulation may not be able to keep track of such a large change, so there will be distortion until it catches up with the signal. This effect, called *slope overload*, is clearly seen in Figure 15. Third, if the signal is not changing, delta modulation cannot represent it, but must instead code it as alternating 0's and 1's, which introduces noise in the playback.

Although delta modulation is an interesting technique, it is not widely used.

ADPCM   With another technique, known as **adaptive differential pulse code modulation,** or **ADPCM,** the encoding circuitry tries to guess the value of the next sample based on values of previous samples (therefore, adaptive). When it takes the next sample, it encodes and stores the difference (therefore, differential) between its prediction and the actual sample. Since the decompression circuit is using the same predictive algorithm, it can re-create the original signal by applying the error correction to each of its generated samples.

ADPCM achieves good compression rates. However, it is more complex to handle than PCM, because of the need to run the predictive algorithm. It is used in personal computers, a recent example being its use to encode the sound tracks used in the ActionMedia II digital video files. The Sound Blaster 16-bit cards support an ADPCM file format that provides a 4-to-1 compression ratio of audio. However, ADPCM format is not standardized, so its use is not yet widespread.

SUMMARY

A characteristic of multimedia data is the enormous size of its files; high data rates are required to deliver the files to the user. Although computers and networks are becoming faster and less expensive, the explosion in multimedia materials dictates that compression is key to enabling the use of these materials in personal computers and networks.

Multimedia data can be successfully compressed using a variety of techniques, each suited to different media. Trade-offs can be made between the amount of compression achieved and an acceptable level of playback quality. The use of compression always involves processing of the data, either by the personal computer's processor or by an auxiliary processor dedicated to the task.

4

Graphics and Animation

Graphics and images lend interest to a document or program, and are an aid to understanding the material being presented. Before launching into a discussion on images, we will discuss some requirements for computer color representation.

Colors

Computer monitors are based on television technology, and display their range of colors by lighting red, green, and blue (the **RGB** system) phosphors on the screen. The colors are additive, so that if a pixel has all three colors lit, the user will perceive the pixel's color as white. Although a program might represent its color to the user employing a different system (hue, saturation, and brightness, or cyan, yellow, magenta, and black, for example), at the hardware level colors are always represented by a combination of red, green, and blue.

Modern color monitors are analog, meaning that the computer can vary the intensity of light for each of the three colors for each pixel from none to some maximum value, thus achieving a rich range of color reproduction capability. However, in the computer itself there is a cost for providing a large number of colors: the amount of memory required, as shown in Table 3.

Note that the memory requirements apply to the display adapter as well as to the program processing the color (e.g., a paint program). If the program does any processing of the image, it will need more memory and processing power to handle more colors. Programs must also be able to run on systems that support fewer colors than the program does. Most programs that handle 16- and 24-bit color will also run on 8-bit color systems.

A technique called *dithering* is used when the number of colors available is less than the number of colors needed to display a picture. Dithering generates patterns of alternating colors that create the illusion of a color that the system

TABLE 3 | *Correspondence of Number of Colors with Memory Requirement per Pixel*

Number of Colors	Bits Required
2	1
4	2
16	4
256	8
65,536	16 (2 bytes)
16,777,216	24 (3 bytes)

cannot display. For example, you can simulate pink by alternating red and white pixels. Dithering is far from perfect, but it lends the illusion of more colors to 16-color systems.

A problem for people who require precise color control is that the color output of monitors varies considerably. Thus a color that looked good on the display may not be acceptable in print. To solve that problem, organizations have defined standard color values that are keyed to color sample charts. The **Pantone Matching System** is a well-known example. Programs that support Pantone colors allow the user to select colors by Pantone name, so that no matter what the color looks like on the screen, it will exactly match the Pantone color sample when printed.

Graphics

Graphics are categorized, depending on the format they are stored in, as either **bitmaps** or **metafiles.**

BITMAPS

Bitmaps are stored as a collection of bits, each bit (or series of bits) representing the color of a pixel. The order of the bytes in the file is directly related to the position of the pixels on the screen. There is no size information stored with a bitmap; thus, a bitmap will appear smaller on a higher-resolution screen. Bitmaps, as a rule, are generated by paint programs, such as the Windows Paintbrush program, Fractal Design Painter, CorelPHOTO-PAINT, and so on.

Bitmaps are well suited for portraying photographic information, such as scenes and people. Because little processing is required when loading a bitmap (unless compression is used), bitmaps load very fast.

Colors Bitmaps are stored in a variety of color "depths." The simplest bitmaps are monochrome bitmaps, in which each bit in the file represents a pixel. Other popular forms include 16-color, 256-color, and 16-million color. Anything less than 256 colors will not approach realistic color rendering.

The size of the file required to store a bitmap is calculated according to Equation 1.

Calculating Bitmap File Size **(EQ 1)**

$$\text{Size in kilobytes} = \frac{\text{Width (pixels)} \times \text{Height (pixels)} \times \dfrac{\text{bits per pixel}}{8 \text{ (bits per byte)}}}{1024 \text{ (bytes per kilobyte)}}$$

256-Color Systems   Although 24-bit color systems are widely available, most of the PCs installed today probably are 256-color (8-bit) systems. These systems can, in many cases, render images quite faithfully, and thus have become popular. However, if your choice were limited to 256 colors, your computer would do a poor job of rendering most photographic images. The secret is that 256-color systems can render thousands or millions of colors but only 256 distinct colors at a time. This is because their memories are limited to 1 byte (8 bits) per pixel. In such systems each image includes a **palette,** which is a table that maps each 8-bit value to a specific color. Thus a pixel with a value of, say, 121 might in one image represent a light shade of green, while in another image it would represent burgundy. When Windows displays an image, the image's palette information is loaded into the color hardware, so that the hardware will know which color to generate for each pixel. Actually, there are less than 256 colors available for the image because Windows assigns 20 color values for the system colors, such as the color of a window's border, title bar, menus, and so on.

Plate I is a picture of a stained glass window at the cathedral of Chartres, France. (Plates I–VI are on inside front cover; Plates VII–X are on inside back cover.) Notice that most of the colors in the image are bright. Plate II shows the palette for this picture. Notice that the palette also contains the bright colors. Contrast this with Plate III, a picture with softer colors, and its palette, Plate IV. The stained glass window is rich in blues, reds, and yellows. Notice that its palette contains an abundance of these colors. The picture of the girl at the table has several shades of pink for the girl's blouse, brownish shades for the girl's skin, and many very dark shades for the background. Notice that the palette for this picture (Plate IV) reflects this different set of colors.

Normally, palette handling is invisible to you. Programs that generate or capture images create the palette information for the image automatically, and programs that use images furnish the image's palette information to Windows automatically when the image is displayed. However, complications arise when you try to display two images at the same time. If the image palettes are different, which palette is used? Software motion videos also present this problem because as the movie progresses and scenes change, it is possible that the palette also needs to change. Although changing the palette is no problem, it will cause unpleasant shifts in color in other windows on the screen. This is called **palette shift.** Consider two images, Plate V, a picture of daffodils, and Plate VII, a picture of a pink azalea. The palettes for these images are shown in Plate VI and Plate VIII, respectively. Neither palette will work well with the other's picture. Plate X shows the pink azalea image displayed using the palette for the daffodils.

To solve this problem, several programs can "optimize" palettes. These programs analyze the images and palettes of each image, or of the frames in a movie, and create a single palette that is the best compromise generated from the separate palettes. This involves finding common colors among the images, discarding, if necessary, colors that are sparsely used, and combining colors that are similar into a single color value, thus reducing the number of palette entries needed for each image. Plate IX shows an optimized palette generated for the daffodils and pink azalea images.

Manipulating Bitmaps   Bitmaps do not lend themselves well to resizing. To shrink a bitmap, the computer program discards pixels from the bitmap, thereby losing picture information. To enlarge a bitmap, the program has to fill in pixels (it cannot scale the pixels themselves), which entails duplicating pixels or interpolating between adjacent pixels. Enlarged bitmaps often have a blocky, jagged appearance. Text represented as a bitmap does not enlarge well.

The pictorial elements in a bitmap do not have any specific representation in the bitmap. Thus in a picture of an automobile, the computer cannot easily isolate such elements as a wheel or a door. Paint programs provide "scissor" tools that allow you to laboriously cut out such features. They also provide "magic wand" tools that select all pixels in the vicinity within a specified range of colors, which in many cases does the job better than the scissor tools. Editing a bitmap is akin to working with a painting on canvas. To change the color of a portion of the image, you change the color of the pixels, which is similar to covering the paint in that area of the canvas with paint of the new color.

File Formats Bitmaps are stored in a large variety of file types, or **file formats.** Sophisticated graphics programs handle several different formats, and other programs, such as Hijaak Pro, specialize in converting between formats. Following are just a few of the popular formats:

- **Tagged image file format (TIFF)** is popular, although there is a proliferation of versions of TIFF, which means that a program that handles TIFF files might not be able to handle your particular TIFF file. A TIFF file has tags to describe its contents. If the program does not understand a particular tag, it just skips it.
- **BMP** and **DIB** (for bitmap and device independent bitmap) are bitmap files in a format defined by Microsoft Windows. They are simple and popular. However, be aware that there are several forms of bitmap files, some not compatible.
- **PCX** is a format defined by ZSoft. This too is a popular format, and it does not suffer from the existence of a number of incompatible versions.
- Kodak **Photo CD,** a newer format, was defined by Kodak for the storage of photographs on CD-ROM. A distinguishing feature of Photo CD is that the format stores each image at a number of resolutions, thereby bypassing the need to scale the images. Low-resolution versions of pictures are appropriate for display on a television set, while higher-resolution versions are appropriate for use with computers and for photo reproduction. Some vendors, such as Corel, have available libraries of photographs in Photo CD format available for purchase.

METAFILES

Metafiles are sequences of drawing instructions that the computer executes to form the image. Drawing with metafiles is also known as *vector-based* graphics. No pixel information is stored in a metafile. The instructions are very similar to, say, drawing instructions in the **BASIC** language. When the computer reads a metafile, it processes each command in the file in sequence, building up the elements of the picture. Whereas bitmaps tend to appear instantly on the screen, with a metafile one might see the individual elements being drawn in sequence.

Metafiles are ideal for simple graphics, diagrams, drawings, and so on. They do not lend themselves well to handling photographs or paintings because the sequence of drawing instructions needed to produce such a picture would be very large, and therefore the file would be quite big and take a long time to draw.

Metafiles are generated by programs such as Excel (to create business graphs), CorelDRAW!, Professional Draw, Micrografx Designer, and Autocad. Figure 16 is an example of a drawing rendered as a metafile. Another example is Figure 7 on page 12.

FIGURE 16   *Picture Drawn from a Metafile*

Manipulating Metafiles   The advantage of metafiles is that they are resolution-independent, because they consist of drawing instructions to the computer. At the time of drawing, the computer will execute the instructions and thus determine which pixels on the screen or on the printer will be affected. For example, a metafile might specify that a circle be drawn in the center of the screen, with a radius of $\frac{1}{8}$ of the width of the screen, with a black outline and a red fill. Running the same metafile on a computer with a high-resolution screen will simply cause the line to look sharper. The only drawback is that pixels on higher-resolution displays are smaller, so lines will tend to look thinner.

Scaling is also not a problem for metafiles. If you stretch a drawing created with a metafile, the computer just changes the scale and draws the picture from scratch.

A big plus for metafiles is that they support the concept of a drawn object. For example, if your picture contains a circle, you can select the circle and delete it. The computer responds by removing the instruction that drew the circle from the metafile; when the drawing is redrawn, the circle no longer appears. Working with metafiles is akin to working with a stack of transparencies, with each object placed on its own layer. If a circle is drawn, and then a rectangle is drawn to cover it, the circle still "exists" as far as the computer program is concerned. If later the rectangle is moved or erased, the circle will again be visible.

A number of graphics programs provide utilities to trace bitmaps, in other words, to create a metafile that duplicates the visual result produced by a bitmap. These are useful because it is a lot easier to manipulate a metafile than a bitmap, as we have already discussed. These programs work by looking for edges (locations where different colors meet) in the image and creating color-filled shapes from the edges. Tracing programs do not work well on complex bitmaps, and they generally require a good deal of manual editing before the metafile produces an image similar to the bitmap's.

Formats   There are several metafile formats in use. Two common formats are:

1. **Windows metafile (WMF).** Most Windows graphics programs support this format even if they, in addition, use another proprietary format.

2. **Computer Graphics Metafile (CGM).** This metafile format was defined as a way for DOS graphics programs to exchange information, and it is still popular with Windows users.

Scalable Fonts   **Scalable fonts** are a special kind of metafile. Earlier graphics environments, such as Windows and the Macintosh, relied on bitmapped **fonts.** This meant that a separate file was required for each point size you wanted to use, as scaling the bitmaps produced unacceptable results. Metafiles were also tried. so that each letter could be generated on the fly based on a set of drawing instructions. However, this was not the answer either, because letters became too skinny as the size increased, and suffered from other visual defects.

To solve these problems, scalable font technology was developed. Letters are drawn according to mathematical formulas (akin to how metafiles work). However, additional information (called *hints*) is also stored that allows the computer to adjust the look of each letter for the specific font size.

Scalable fonts have proven to be very popular and thus are supported on all platforms. The two best-known scalable font technologies are Adobe's Type 1 (used in Postscript and supported in Windows through the use of the Adobe Type Manager program) and TrueType (with support built into Windows). A large variety of fonts for these technologies are available for purchase. Many programs, such as word processors, graphics programs, and desktop publishing packages, include a selection of such fonts. CorelDRAW! 5.0 includes over 800 fonts in both Adobe Type 1 and TrueType formats.

A newer technology from Adobe, available in the latest version of Adobe Type Manager, is called *multiple masters*. Multiple masters allow you to smoothly change a characteristic of a font along an "axis." For example, instead of having to choose between either a normal font or a bold font of a **typeface,** a multiple master allows you to generate a font from the typeface with any degree of "boldness" in between by varying the weight axis of the font. Figure 17 shows the effect of the "optical" axis on Adobe's Minion Multiple Master typeface. The top word was rendered in 72 point with a font designed for 72 point. The bottom word was rendered in 72 point with a font designed for 6 point. Notice that the font designed for 6 point has thicker stems and *serifs,* and looser fit. Thus, when rendered at 6 point, the 6-point font will be more legible because the stems will not be too thin nor the letters too close together. This capability is important for those in the publishing business, as they can vary a typeface until it looks just right for the particular circumstance.

Do not confuse fonts and typefaces when purchasing a font package. Each variation of a typeface is a font. So, having normal, bold, italic, and bold italic versions of a typeface means that you have four fonts but only one typeface. Some popular typefaces may be available as ten or more fonts, with variations such as bold, semibold, black, light, condensed, and so on.

Printing Graphics

One would think that printing a graphic is no more difficult than placing it on the monitor screen. However, printing is a different technology, so there is much to consider when trying to print images and graphics.

MONOCHROME PRINTING

Consider a simple example, such as a circle filled with a light color. Let's say we want to print it on a laser printer. On the screen we can reproduce many colors

FIGURE 17 *The effect of the optical axis on Adobe's Minion Multiple Master typeface.*

Hamlet

Hamlet

and shades by varying how much we light each red, green, and blue pixel of the picture. The more bits we have for each color, the more finely we can control the shade. With 24-bit color, we can light each of the red, green, and blue components of the pixel at any of 256 levels. With current printer technology, however, you only have two choices: either to print a dot or not print it; there is no way to control how light or dark the dot is. Furthermore, all dots are the same size. Thus the program must simulate the shade by printing some dots and not others. If the shade is light, we will end up with widely spaced dots (Figure 18), which do not look too much like a light shade.

Another problem is that the screen has a lower resolution than the printer. A screen will have perhaps 72 pixels per inch, while most laser printers can print either 300 or 600 dots per inch. A picture that might look fine on the screen will appear too small when printed. One solution is to define larger dots, each one composed of a number of printer dots. By varying the size of the composite dots you can simulate shading. This technique, called *halftoning,* is used in newspaper photographs. If you look at one closely you will notice this. The dots, although varying in size, are equally spaced. The number of dots or lines per inch is called the *screen frequency.* You can also control the alignment of the dots, which is

FIGURE 18 *Circle with Light Shading*

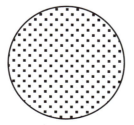

called the *screen angle*. High-end graphics programs, such as CorelDRAW!, allow you to control the screen frequency and the screen angle when printing a picture.

COLOR PRINTING

There are two main distinctions between color on the screen and color as printed on paper:

1. Color on a computer or television screen is called *additive* color, which means that red, green, and blue light is added together to give the illusion of a color. On paper, color is achieved by a *subtractive* process, and the colors used are cyan, magenta, and yellow. For example, when there is yellow ink on the paper, the paper appears yellow because the ink absorbs (subtracts) most of the colors of the white light but reflects the yellow light.
2. Most printers and printing presses can only print one color at a time, so to print a multiple-color picture it is necessary to print the page several times, laying one color on top of another.

Process Color   *Process color* printing consists of printing color pictures using the primary colors cyan, magenta, yellow, and black, also known as **CMYK.** Theoretically, combining cyan, magenta, and yellow should produce black, but in practice it produces a muddy dark color, so black is used to produce crisp blacks. Another reason to use black is that printing presses do not perform well if they have to lay a lot of ink on the paper. Producing a dark color by laying on a high saturation of three colors of ink is more than the presses can handle. Thus a technique called *black removal* is used, to remove any black component of the color from each of the three primary colors and add it to the black ink. Here is how it works: Suppose a fill calls for 60% cyan, 60% magenta, and 20% yellow (twilight blue). The black component is made up of equal amounts of each primary color, so if we subtract as much of the primary colors as we can, in equal amounts from each color, we will end up with 40% cyan, 40% magenta, 0% yellow, and 20% black. The 20% black is equivalent to 20% cyan, 20% magenta, and 20% yellow. Before we did black removal we had total ink quantities of 60%, 60%, and 20%, which equals 140%. (Remember that the printing takes place in three separate passes, so the maximum total is 300%, not 100%.) By doing black removal we have 40% cyan, 40% magenta, and 20% black, for a total of 100%, meaning we need less ink on the paper to obtain the same color.

High-end graphics programs and desktop publishing programs handle black removal. To create the plates needed for printing, these programs can produce color separations. A color separation is the process of printing, in monochrome, a sheet with only the cyan components of the image, a sheet with the magenta components of the image, and so on. For example, assume that a portion of a picture consists of the color called "grass green", whose components are 60% cyan, 0% magenta, 40% yellow, and 40% black. Assuming the separations are printed on a laser printer, the area corresponding to the grass-green color on the cyan separation will be 60% saturated (a darker grey), the magenta separation will have 0% saturation (white), and both the yellow and the black separations will have 40% saturation (a lighter shade of gray). From these four separations, four photographic plates are prepared, and from there, the four press plates. When the paper goes through the press, the first plate will lay on cyan ink with 60% saturation, the second plate will not lay on any magenta in that area, and the third and fourth plates will lay on yellow and black ink with 40% saturation, with the resulting color being grass green.

Sometimes, when the print process is of low quality (e.g., Sunday newspaper comics) the plates are not correctly registered so they print slightly offset from one another. In these cases you can clearly see, especially around the edges, the separate colors that make up process color.

Spot Color   Sometimes you want to make absolutely sure that what you print matches a color exactly. Or, you might have only two colors in the publication. It costs less to print fewer colors, so in this case preference is given to what is called *spot color*. With spot color, you specify, using a system such as Pantone, the exact color that will be printed by each plate. Of course, the spot color could be broken down into its components and printed using the CMYK model (process color), but the color will be better and use less ink using spot color.

What if you have a picture with, say, 10 colors? Your only choice then is process color, because spot color would require 10 separate printing plates. This would be very expensive, and would probably lay too much ink on the paper using 10 separate printing steps (not to mention the fact that 10-color presses simply do not exist).

There is enough material on the topic of printing to fill a book by itself. We have just lightly covered some of the highlights.

Animation

Animation consists of playing a succession of graphic images on the screen, either by displaying a sequence of stills in quick succession or by continuously erasing and redrawing certain visual elements on the screen (e.g., a drawing of a car) at different positions.

The most primitive type of animation is one in which each frame, also called a *cel*, is drawn individually. This technique was used to create all of the Disney classic animation films. Each cel was individually colored. "Primitive" here means that the technique is very labor intensive and not aided by computer programs; it does not refer to the results, which can be spectacular if enough talent and effort are applied to the process.

ANIMATION PROGRAMS

Animation programs allow you to more easily create animations, which can then be incorporated into other programs. The two prevalent formats for animation files are **FLI and FLC** (called *flick*), which use a format defined by Autodesk, and the **MMM** format defined by Macromedia. Animation programs use a theater metaphor for the creation of an animation. Thus the space in which you create the animation is called the *stage,* which contains the fixed elements, sometimes called *props* (e.g., the background). Props are shown in Figure 19. *Actors* are the objects to be animated. In an animation of a car passing by, the scenery and the road are the props on the stage, while the image of the car is the actor.

Path-based animation is very popular. In this method you specify (draw) a path for the actor to follow, and the animation program will create each frame, with the actor in successive positions along the path in each frame. You can specify the time the actor takes to travel the path, or the number of steps the trip should take. You may also specify key positions on the stage, and have the program fill in the missing steps through a process called *in-betweening.*

These animations are adequate for some purposes, such as showing the trajectory of a ball through the air. However, it is often necessary to change the appearance of the actor during the animation independently of the path of the animation. An animation of a person walking consists of not only the movement of the

FIGURE 19   *Props for an Animation*

actor along the path but also of different views of the actor, to simulate leg and arm movement. Thus the actor might consist of a number of images played in succession as the actor travels along the path. In this case the actor is called a *multiple-cel* actor. A multiple-cel actor consists of a small number of related images showing the different phases of movement, such as a person walking, a bird's wings flapping, a horse galloping. To create the animation the program moves the actor along the animation path displaying successive images of the actor on successive frames of the animation. Suppose the actor has three cels (three images). In creating the animation, the program does the following:

- On the first frame, the first cel of the actor is shown at the starting point.
- On the second frame, the second cel of the actor is shown at the second point along the animation path.
- On the third frame, the third cel of the actor is shown at the third point along the animation path.
- On the fourth frame, the first cel of the actor is shown at the fourth point along the animation path, and so on.

Figure 20 shows a set of nine cels that can be used to create an animation of a dinosaur walking. The cels are an example of animation clipart provided with the CorelMOVE program. The dinosaur is a multiple-cel actor. Playing the cels in sequence will create the illusion of walking, but the dinosaur will appear to walk in place. By combining cel and path animation, however, a realistic animation is obtained. Figure 21 shows part of a screen of the CorelMOVE animation program. The prop of Figure 19 forms the background, and the actor of Figure 20 (surrounded by a dotted line) will be animated. The path of the animation, shown as the solid line, is specified by clicking the mouse at each sequential point you wish the animation to follow. Each little square along the path represents a point on the path, in other words, a place where the mouse was clicked. When played, the animation shows the dinosaur strolling through the park. Notice also the Path Edit

FIGURE 20   *A Set of Nine Cels for an Animation of a Dinosaur Walking*

tools, which allow you to edit the path, smooth it, add or delete points, and arrange the points so that they are equally spaced. When the animation is played (with the VCR-like controls at the bottom of the picture), the animation program will show the next cel at the next point on the path, then the cel after that at the following point on the path, and so on. After the ninth cel is shown, the first cel is shown again.

Once the animation is the way you want it, you can create an animation file that will run within another program. Windows itself contains the animation driver, which the other program will invoke whenever you choose to run the animation.

As a rule the actor will have many fewer cels (images) than the number of frames comprising the animation. Consider another example: A bird will have only enough images to illustrate one cycle of wing-flapping. By playing the sequence of images over and over along the path, you can create a convincing animation of the bird flying across the screen, such flight consisting of many wing-flapping cycles—in other words, many repetitions of the complete sequence of cels comprising the multiple-cel actor.

Animation programs include paint tools that enable you to create actors and stages. Most animation programs also provide clip-art actors and allow you to bring in actors and props from other programs. Animations may be exported in a standard file format so that the animation can be played on a computer that does not contain the animation program. An increasingly popular option is to export the animation as a software motion video in Video for Windows or QuickTime for-

FIGURE 21   *Specifying the Animation Path*

mat. This allows the animation to be edited and incorporated into other movies using such programs as Adobe Première.

These programs can animate a number of objects (actors) at the same time. The process of creating the animation might be slow, as the program must calculate the position of each object for each frame and handle the succession of images for each multiple-cel actor. However, animations can play back at a realistic speed, because during the playback process no computing is taking place. The computer is simply reading each successive frame and placing it on the screen. The individual frames of an animation may be compressed, but the compression scheme is simple (e.g., run-length encoding), so decompressing each frame is a simple process.

There are also programs that create and use 3D animation. In 3D animation, a mathematical model of the object is used to compute the view of the object at each point in time. This view depends not only on the size and shape of the object but also on its position relative to the viewer. Sophisticated 3D modeling programs also take into account the lighting of the environment as well as the texture of the object. Because of the number and complexity of calculations involved, such animations are done mostly on **UNIX** workstations running programs such as Wavefront. However, one program that employs 3D animation has been with us since the early days of the personal computer: Microsoft Flight Simulator. In the latest version, buildings and other objects have shadows that vary according to the time of day and the latitude and longitude of the observer.

Examples of animation programs include Autodesk Animator Pro, Macromedia Director, and Gold Disk's Animation Works Interactive. The CorelDRAW! package includes the CorelMOVE animation program, which was used to create the dinosaur animation discussed earlier.

PROGRAMS INCORPORATING ANIMATION

Many programs have built-in animation features. With these programs, animation consists of moving program-defined objects around on the screen. The animations produced by these programs are specific to the programs and cannot be exported for use in other programs. The animations are stored as a part of the programs' data files instead of in animation-specific files. Some examples include:

- **Asymetrix Compel.** This is a program intended for the production and delivery of on-screen presentations. Bulleted text can be made to appear from any position on the screen and move into place. Other objects can also be made to move along a specified path. In addition, this program supports the playing of animation files produced by other programs.
- **Macromedia Action.** This program provides function and capabilities similar to those in Asymetrix Compel.
- **Authorware Professional.** This is a program intended for the creation of courseware. It provides a rich set of animation capabilities. Objects can be made to move along a straight or curved path. The speed of the animation can be controlled, and the animation can even proceed in step with the changing values of variables, making this a good program for simulations.
- **Asymetrix ToolBook and Multimedia ToolBook.** These are general-purpose hypermedia authoring programs. They do not include specific animation capabilities. However, the positions of objects in the programs can be controlled, and thus animation can be achieved by moving one or more objects in small steps. Multimedia ToolBook 3.0 includes an animation generator that supports path-based animation and multiple-cel actors. Using this feature produces smoother animations than those obtained using the ToolBook programming language.
- **Microsoft Visual Basic.** This is a general-purpose programming system. Certain elements in a Visual Basic screen can be animated in the same way as ToolBook objects.
- **Interactive Physics and Working Model.** These programs perform accurate simulations of two-dimensional mechanical systems. The user of the program builds a physical apparatus out of the program-supplied objects (blocks, pulleys, ropes, pistons, springs, etc.); the program then carries out a simulation of the system, using animation to illustrate the behavior of the objects.

This is but a small sample of programs employing animation. Animation is not limited to fun and games. It is used to illustrate physical concepts, a sequence of actions, and so on. A process might be too complex to illustrate it with a movie; simplified animation might be a much better approach to explaining the concepts.

5

CD-ROM

Compact Disc Read Only Memory, also known as CD-ROM (Figure 22), is a significant technology that affects not only multimedia but the delivery of programs and data as well. CD-ROM is closely related to consumer-oriented compact discs, a technology that has rendered the vinyl phonograph record obsolete.

CD-ROMs, though slow to catch on, have become a significant contributor to the success of multimedia on personal computers. CD-ROM discs allow software developers to include sufficient multimedia content, in the form of pictures, video clips, and audio files, to make their titles comprehensive and meaningful. Wide-scale adoption of CD-ROM discs as a software delivery medium has been driven by the dramatic decrease in CD-ROM player prices, the increasing performance of personal computers (which makes playing multimedia feasible), and the popularity of CD-ROM–based entertainment (games) and consumer-oriented products. Some CD-ROM game titles have sold several hundred thousand copies. Other popular CD-ROM topics include encyclopedias, medical information programs, atlases and travel programs, and so on.

In this Chapter we will examine CD-ROM technology and applications in general. In Chapter 6, we will cover audio applications of CD-ROMs.

Introduction to CD-ROMs

The CD-ROM is a **read-only** medium that stores information in digital format, much like a disk or diskette. The difference is that disks and diskettes are magnetic media, so the information is stored as changes in the magnetic field. CD-ROMs are optical devices. The bits are stored in the form of microscopic pits in a reflective material (Figure 23). The CD-ROM drive contains a laser that shines on the disc as it spins. As the pits pass by, the amount of laser light reflected changes, which the drive indirectly interprets as 1's and 0's. The CD-ROM track is laid out as a spiral, starting at the center of the disc and moving outward. The

FIGURE 22   *CD-ROM Disc*

FIGURE 23   *Diagram of a CD-ROM Disc (not to scale)*

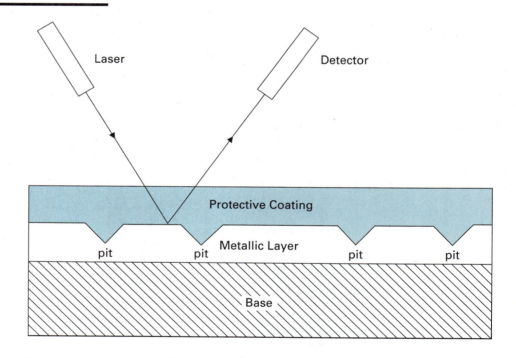

drive varies the speed of the disc (how fast the disc spins) depending on where on the disc the laser is shining.

CD-ROM discs are very portable and durable. Scratches made in the protective plastic material that encloses the reflective surface generally do not affect the readability of the CD-ROM.

CD-ROMs are direct descendants of the compact disc, developed for the consumer market by Sony and Philips. Most computer CD-ROM drives provide audio output jacks so that you can listen to compact discs using the drive. It is also possible to mix audio and computer files on the same CD-ROM. One example is Microsoft's Multimedia Beethoven CD-ROM, which contains a rendition of Beethoven's Ninth Symphony, along with a computer program that guides the user through the composition.

Computer specialists sometimes use terms such as "Red Book audio" or "Yellow Book" when discussing compact discs and CD-ROMs. These terms refer to the color of the covers of the specification books for the various kinds of compact disc products. Red Book refers to the compact disc standard. Yellow Book is the standard that defines CD-ROM. The Yellow Book specification modifies the Red Book specification by adding a table of contents to the disc, so that the computer can locate the files on the disc. It also strengthens the error detection and correction capabilities of the player, so that there is a high degree of confidence that the data can be reliably retrieved from the disc, even if the disc is somewhat scratched or smudged.

There are derivatives of the Yellow Book standard. The Macintosh HFS (Hierarchical File System) format is defined for Apple's Macintosh computers. A more universal standard is ISO (International Standards Organization) 9660, which is the standard implemented on IBM-compatible personal computers. The ISO 9660 driver is also available for the Macintosh, meaning that the same CD-ROM disc, if it conforms to the ISO 9660 standard, can be read on both Macintosh and Windows platforms. If the CD-ROM disc contains software, however, the software must be available in both Macintosh and **DOS**/Windows versions, as software intended for one platform will not run on the other.

There is also a Green Book that defines the standards for CD-I, Compact Disk Interactive, a consumer-oriented technology developed by Philips. There are standards with other color names as well.

The Importance of CD-ROMs

In addition to music, CD-ROMs may also contain data files, perhaps some image and audio files, and computer programs. The amount of information that can be stored on a CD-ROM is impressive. If you were to unwind the pits and spaces from a CD-ROM, the trail would stretch for three miles.

A full-size encyclopedia, including pictures, easily fits on a CD-ROM, as does an entire shelf of reference manuals or the twenty-volume *Oxford English Dictionary*. Four CD-ROMs, which take up minimal space on your desk, can hold approximately one million pages of text.

CD-ROMs can provide efficient storage of material such as:

- On-line publications such as catalogs and technical manuals
- Classroom programs
- Resource libraries
- Graphics libraries, clip-art illustrations, and music libraries
- On-line specialized periodicals (legal, medical, and archival)
- **Photo CD** high-resolution image collections

A number of vendors are now shipping their software programs on CD-ROM. This reduces the manufacturing costs for the vendors and makes installation of the software much easier for the user, as there is no need to handle multiple diskettes. The storage capacity of a CD-ROM also allows vendors to add extra bonuses with their programs, such as clip-art collections, on-line manuals, extra sample programs, and so on. Examples of products on CD-ROM include CorelDRAW!, Authorware Professional, Multimedia ToolBook, Microsoft Encarta,

The Oxford English Dictionary Second Edition on Compact Disc, McGraw Hill's Multimedia Encyclopedia of Mammalian Biology, and DeLorme's Street Atlas USA.

The Oxford English Dictionary Second Edition on Compact Disc is a good illustration of the impact of CD-ROM technology. The hardcover version of the dictionary spans twenty large volumes, and costs several thousand dollars. The CD-ROM version contains the full text of the hardcover edition on a single disc and retails for less than one thousand dollars. In addition, the CD-ROM product provides a number of search capabilities that allow you to very quickly locate a word or words using such diverse criteria as etymology, part of speech, definition, and so on. These searches would be impractical, if not impossible, to conduct using the hardcover edition.

CD-ROM Characteristics

Think of a CD-ROM as just another storage device on your computer, with the following characteristics:

- Massive storage capacity. The capacity of a CD-ROM exceeds 600 megabytes. How much is used by a particular CD-ROM will depend on what the author included on it. The file organization is exactly the same as on a hard disk.
- Read only. Like a compact disc, a CD-ROM is a read-only medium. You cannot use it to store documents and programs you have created. This is an advantage in that you cannot accidently write over a program or other file contained on CD-ROM, something that occasionally happens with files on disks and diskettes.
- Removable. You probably will need only a single CD-ROM drive, even if you have multiple CD-ROMs. In this respect they are very much like diskette drives. Some CD-ROM drives require the use of a caddy, an enclosure into which you must place the CD-ROM before inserting it in the drive. Caddies can be purchased at most computer and software retailers.

Two other factors related to CD-ROMs that are important to take into account are the **access times** and **transfer rates** supported by CD-ROM drives. *Access time* refers to the time that lapses from when the CD-ROM drive receives a command to retrieve some data until the data starts flowing to the computer. CD-ROM drives are much slower than hard disks in this respect, partially because the CD-ROM track is laid out as a spiral; the drive must seek the approximate location of the track, and then read the CD-ROM to locate the specific information. Today's fastest CD-ROM drives have average access times of under 200 milliseconds (hard disks can have average access times as low as 10 milliseconds). The shorter the access time, the faster the data will start flowing.

Most CD-ROM drives support a *transfer rate* of slightly more than 150 kilobytes (150,000 bytes) per second, which is the transfer rate needed to support the playing of audio compact discs. Unfortunately, this is rather slow by computer standards. Some software vendors who ship their software on CD-ROM encourage users to copy the program to the hard disk so that it will load and run faster. A newer development is the *multispeed* (sometimes called **multispin**) CD-ROM drive. When not playing compact disc music, these drives double, triple, or quadruple the speed at which they spin the CD-ROM, thus achieving transfer rates of up to 612 kilobytes per second (e.g., Pioneer DRM-64X). The most recent development is an NEC CD-ROM drive that can spin the CD-ROM disc at six times the standard rate; that is, it transfers data at over 900 kilobytes per second. Most CD-ROM drives shipped today are multispeed drives. Unlike phonograph records, a CD-ROM disc can be played at any speed. The only difference is that on a multispin CD-ROM drive the data is retrieved at a higher rate (faster). The data cannot come off the disc too fast, because the computer is capable of receiving the data of much faster devices, such as a hard disk. When a multispin CD-ROM drive is playing a conventional compact disc, or Red Book audio on a CD-ROM disc, the drive will automatically slow down to the correct speed.

CD-ROMs are key to multimedia because they can easily store the large files multimedia demands. For this reason, the Multimedia Personal Computer Level 2 Specification (a specification detailing the minimum requirements for a personal computer to be multimedia capable) defines that a CD-ROM drive's average access time should not exceed 400 milliseconds and its transfer rate should be at least 300 kilobytes per second.

Another important consideration for those buying aftermarket CD-ROM drives is the method used to attach the drive to the computer. Most CD-ROM drives have an **SCSI** (small computer systems interface, pronounced "skuzzy") interface. Many of today's personal computers also provide a SCSI interface, so theoretically they should connect with no problem. However, variations in implementation dictate that you should "try before you buy." CD-ROM drives using other interfaces come with an adapter **card** that plugs into the computer. CD-ROM drives also require special drivers, which are programs that translate computer commands into CD-ROM commands. The most common driver is Microsoft's **MSCDEX.** The drivers you need should be furnished with the drive or by the retailer who sells the drive.

CD-ROM XA AND MULTISESSION CD-ROM DRIVES

What if you had some digital audio on a CD-ROM that you wanted to play as background for a slide show that is also on the CD-ROM? Given the slow access time of the CD-ROM, and the size of the files, the music would stop each time the drive went after the next image. CD-ROM **XA** (the XA stands for "extended architecture") was defined so that data from different files can be interleaved on the CD-ROM, meaning that the CD-ROM track contains a number of bytes from the first file, followed by a number of bytes from the second file, followed by more bytes from the first file, then bytes from the second file, and so on. Thus the two files can be read without having to move the CD-ROM drive head back and forth between two tracks, which is very time consuming. XA is also a requirement for Photo CD access. Most of the CD-ROM drives shipping nowadays are XA drives.

Multisession is also a requirement for Photo CD compatibility. Kodak defined the Photo CD format so that consumers could add photographs to the disc at various times (such as when they have another roll of film developed). Earlier CD-ROM drives could only access the photographs stored during the first session (the time the first photographs were written to the disc). This is not a problem for most commercial Photo CDs, which are written in a single session, but it is a problem for consumer-created CD Photo discs. Most, if not all, drives being shipped today are multisession capable.

Networking

Because CD-ROM drives are very much like disk drives, they are easily shared over a **network.** The restrictions, in addition to the CD-ROM's read-only characteristic, are related to the access times and transfer rates. In addition, many CD-ROMs contain programs or other copyrighted material, so licenses must be in force to allow multiple users to access the same materials.

Several CD-ROM drives are available with changer or carousel mechanisms that allow one drive to have access to multiple CD-ROMs without human intervention. These are used mostly in libraries where some reference materials, such as collections of magazine and journal article abstracts, span multiple CD-ROMs.

Vendors also sell CD-ROM servers, which consist of towers of CD-ROM drives connected to a controller so that multiple users may share multiple CD-ROMs at the same time. St. Philips College in San Antonio, Texas, has a system with about one hundred CD-ROM drives, allowing multiple students simultaneous access to popular CD-ROMs, as well as providing the school with the capability of placing multiple titles on-line from a central location.

Such CD-ROM servers are very attractive to the higher education community because they cost less than the alternative cost of providing every machine on the network with its own drive. Another advantage is that the CD-ROMs themselves do not have to be given out to students in order to be run, thus diminishing the damage and loss potential.

Kodak Photo CD

The Photo CD was initially aimed at the consumer market as a new means of storing photographs. The consumer was to take pictures using conventional film. When developed, instead of receiving an envelope with prints, the consumer would instead receive a Photo CD containing his or her pictures. Due to the necessity of owning a Photo CD player to view the photographs (which were to be shown on a TV set) and the higher cost of the disc over conventional prints, this technology never caught on. However, it is very popular with professional and commercial photographers, libraries, academics, and so on. The advantages of the Photo CD include reasonable cost, durability, compact storage, and the ability of the photographer to edit the picture on a computer.

Photo CDs store pictures in 24-bit color (the pictures start out at 32-bit color), in a compressed, proprietary format. Each picture is present on the Photo CD in 5 separate resolutions, as follows:

- Base over 16   128 lines × 192 pixels
- Base over 4   256 lines × 384 pixels
- Base   512 lines × 768 pixels
- 4 times Base   1024 lines × 1536 pixels
- 16 times Base   2048 lines × 3072 pixels

If the need is for a thumbnail of the picture, a lower-resolution version of the photo is used. For computer applications, one of the medium resolutions will suffice (the standard computer screen resolution is 640 × 480 pixels). To make a high-quality enlargement or poster of the picture, the high-resolution version is used. This format allows a Photo CD to store about 100 pictures. Pictures can be recorded on the Photo CD at different times (for example, each time another roll of film is sent in for processing) until the Photo CD is filled. This implies that a drive must be multi-session capable in order to be able to retrieve all of the photographs on the Photo CD, as the first table of contents on the disc will not be aware of pictures written in subsequent sessions.

Other forms of the Photo CD should become available to meet different needs. Kodak has defined a "professional" resolution, 64 times Base, with a resolution of 4096 lines × 6144 pixels. Kodak also promises formats suitable for medical purposes, to store x rays and other imaging data.

Although the format is proprietary, the driver software (the software that allows the computer to find and retrieve the pictures on the Photo CD) is widely licensed by Kodak. Many image-editing and authoring programs have the capability of importing pictures in the Photo CD format. Some companies use the Photo CD as a medium for distributing photographic clip art, in other words, collections of pictures that can be used in your documents and programs.

A Photo CD, like any other compact disc, can be damaged by improper handling. Discs may break, crack, warp, or become scratched enough that the information on the disc may no longer be recoverable. Given proper handling, Photo CDs and other compact discs should last for more than 100 years. Other than through improper handling, the only way that a disc can be damaged is if oxygen gets through the protective plastic layers and oxidizes the aluminum, destroying its reflectivity. There have been no reports of this phenomena occurring, however.

The Photo CD provides a cost-effective way for you to incorporate your photo or slide collections into computer-based presentations and courseware. It is also a way for you to bring new photographic material to the computer. The same camera and film that are currently used to take photos and slides can be employed to take pictures intended to illustrate documents or programs. This is a more reasonable alternative to the use of most digital cameras, which can feed photographic information directly into the computer. Digital cameras are very expensive, in the range of thousands, or tens of thousands, of dollars.

CD-R

Recordable CD, or **CD-R** is a technology that allows you to record your own CD-ROM discs. While the discs are physically different from prerecorded CD-ROM discs, once recorded they can be played on any CD-ROM drive. Conventional CD-ROM discs are pressed from a master. A CD-R disc has a gold layer covered by a green dye layer. During the recording process, a laser disintegrates the dye layer in spots, allowing the gold to show through, thus achieving the same sequence of reflective and nonreflective areas as in a conventional CD-ROM disc. CD-R discs can be readily recognized by their gold color. CD-R technology is governed by the Orange Book specification.

Creating Your Own CD-ROM

There are a number of reasons why you might want to create your own CD-ROM. You might be creating an educational multimedia program for sale, or you might be attracted by the CD-ROM's potential to store massive amounts of information. Perhaps you have a slide collection you would like to transfer to a more durable medium. You might want to consolidate a set of handouts, reference materials, and tutorials for a course onto a single disc. Whatever the reason, there are two ways of creating your own CD-ROM disc.

The first way is to send your files to a manufacturer who will then create a master from which an almost unlimited number of discs may be pressed. Typically the charge for creating the master ranges from $500 to $1,000. You will, before the process is complete, be able to obtain a check disc which you can use to verify that all the data and programs are as they should be. This is very important if your disc contains setup programs to install other software. After the master is completed, you may purchase any number of discs. Prices will depend on the volume of discs produced. It takes a very large production run to get the cost down to the $1 to $2 range. Sometimes the price of the master includes the cost of a small production run. You will also have to purchase some form of packaging for the discs, such as jewel cases.

The second way to produce a CD-ROM is to purchase a CD-R recording unit that attaches to your personal computer. The recorder creates CD-ROM discs one at a time, by copying files from a hard disk to a blank CD-R disc. This is the preferred method if you only need a few copies of each CD-ROM disc, as would be the case if you were giving one disc to each student in a twenty-student class. Each disc takes from 30 to 60 minutes to record, and requires the use of a fast computer and a large, fast hard disk (some recording units come with their own hard disk). Any interruption during the writing process will ruin the disc being written. Some of these units cost less than $5,000, and prices are dropping. Such a unit would be a good investment for a department that plans to create a number of CD titles for limited distribution, as one can very quickly recover the cost of the unit given the alternative of making masters at $1,000 each. In addition to the cost of the unit, which normally comes with its own mastering software, you should take into account the cost of the blank discs, currently $15 to $30.

SOME USEFUL HINTS

There are a few rules to observe when creating a CD-ROM. If you choose to have a master made, make sure you can deliver the files in the media the vendor requires. Special programs are available to interleave files in XA format.

The file format of the CD-ROM is the same as that on a hard disk. Files are organized in a directory structure and use DOS naming conventions. To create a CD-ROM, you normally write all of the files in the exact order you want them on the CD-ROM onto a large (1 gigabyte) dedicated hard disk. The files are then copied to the CD-ROM disc in order.

The standard for recording CD-ROMs imposes a limit on how deeply directories may be nested (i.e., how many levels of file subdirectories there are). Moreover, long file names (such as those used by OS/2, the Macintosh, and Windows NT) and case-sensitive file names are not allowed. The main problem, though, is with the physical placement of files, especially with time-dependent multimedia materials, such as waveform audio and digital video. Files placed on the inside tracks of the disc will be accessed faster than those on the outer tracks, so critical files should be placed there. Some CD-ROM–mastering software packages include a program to simulate the performance of the CD-ROM, so you can get an idea of performance problems before the CD-ROM is cut.

Finally, consider duplicating frequently used files to cut down on access time. For example, place a copy of the picture-viewer program in each directory containing pictures to be viewed. The duplication of programs will probably not be a problem given the amount of space available.

Disk versus Disc

Just a bit of trivia to close out the chapter. In reading about CD-ROM technology you will find the words "disc" and "disk." Many people use them interchangeably, but actually there is a distinction: Disk (with a *k*) is used to refer to magnetic media, such as a hard disk drive. Disc (with a *c*) is used to refer to optical media, such as compact discs, laserdiscs, CD-ROM discs, read-write optical discs, and so on.

6

Audio

The applications for audio on personal computers go well beyond producing special sound effects for computer games. Consider the following:

- A music student uses the computer to control the playing of a symphony. Difficult passages can be repeated at the touch of a key, and the computer provides written commentary on-screen synchronized to the music.
- A foreign-language student uses the computer to listen to the pronunciation of words spoken by natives in the language being studied. The student can click on a written word on the screen and hear it pronounced.
- A zoology student uses the computer to hear the different vocal sounds made by different species of the animal they are studying.
- A medical student uses the computer to study the sounds produced by different kinds of heart defects.

The list is almost endless. In this chapter we will discuss the three principal technologies employed to produce audio under the control of the computer, which are:

1. CD-ROM Audio
2. Digital Audio
3. MIDI (musical instrument digital interface)

CD-ROM Audio

We examined CD-ROM technology in detail in Chapter 5. We also saw that the CD-ROM can be used to store computer files. It can, in addition, be the source of high-quality audio, in other words, serve as a compact disc.

Philips and Sony developed the compact disc, in which music is stored in digital format (see "Sampling" on page 8), in order to create a medium that enabled high-quality music reproduction. The compact disc revolutionized the recording industry, replacing the decades-old technology of phonograph records. Compact discs are small, durable, and not subject to wear by the playback process.

A compact disc can contain up to 72 minutes of stereo audio. This audio is recorded using 16-bit samples (65,536 distinct levels) at a sampling rate of 44.1 KHz, which produces a frequency response of over 20 KHz. Data is stored in PCM format (see page 10) without compression. The theoretical signal-to-noise ratio of compact discs is in the order of 98 decibels. In other words, the ratio of the loudest representable sound to the audible noise introduced by the sampling process is 77,200 to 1. This may seem like overkill, but humans have an enormous hearing range. The sound of a full symphony orchestra playing is 10,000 times louder than the sound produced by a lone piccolo.

Every CD-ROM drive is capable of playing regular compact discs. It plays the discs in response to commands sent to it from the computer. However, unlike the reading of a data file, the musical information read when the compact disc is playing is not sent to the computer. Instead, it is converted into a stereo analog signal, which is available at the output jacks present on the player. These signals can then be fed to an amplifier, powered speakers, and so on.

The computer is able to command the CD-ROM drive to play, seek, stop, pause, eject the disc, and so on. It can specify start- and stop-playing points on the disc as precise as one frame, which is 1/75 of a second (not all CD-ROM drives support resolution to this level, though). Thus the author of a multimedia program has precise control of what music plays and exactly when and where it starts and stops.

The commands necessary to control the CD-ROM drive for the purpose of playing compact disc music are very simple, so issuing these commands imposes no burden on the computer. In other words, it does not require a fast computer in order to use CD-ROM audio for multimedia applications. The fact that the sound signal never enters the computer is also significant in not placing a processing burden on the computer.

The key consideration for the use of CD-ROM audio is copyright. As a rule, you cannot purchase a compact disc and then use it in a multimedia application without obtaining additional rights to the material, although educators are given freer rein (when the materials are used for educational purposes) than most others.

Waveform Audio (Digital Audio)

If you have ever heard the sounds that greet you when you start a Windows computer then you have heard waveform audio. Waveform audio is digitized sound (see "Sampling" on page 8) stored as a data file on a computer's hard disk, on a CD-ROM disc, or that is made available over a network. The sound is played through an **audio card,** or using audio circuitry built in to the computer. It is important to distinguish waveform audio from CD-ROM audio. Although waveform audio can be stored on a CD-ROM disc, it cannot be played through the CD-ROM drive, as compact disc audio can. To the CD-ROM drive, the waveform audio file is indistinguishable from any other data file on the disc.

Waveform audio is normally stored in PCM format (see page 10), although other formats such as ADPCM (see page 22) are sometimes used. Table 4 shows the most commonly used sample rates and sizes.

The sample rate affects the frequency response of the sound. If an application needs to play files containing the pronunciation of words, then voice-quality sound is all that is needed.

The sample size affects the noise heard on playback. Sample files of 8 bits are audibly noisier than 16-bit sample files, so for high-quality audio, 16-bit samples are used.

TABLE 4 *Waveform Audio Types*

Sample Rate (KHz)	Sample Size (bits)	Frequency Response	Data Rate: Mono	Data Rate: Stereo	Quality Measure
11.025	8	5 KHz	11 KB/sec	22 KB/sec	Voice
22.05	8	10 KHz	22 KB/sec	44 KB/sec	FM radio
22.05	16	10 KHz	44 KB/sec	88 KB/sec	FM radio
44.1	16	20 KHz	88 KB/sec	176 KB/sec	Compact disc

There are a number of popular sound cards on the market. Perhaps the leader is the Creative Labs Sound Blaster series of cards. Others include the IBM M-Audio card and the new IBM Windsurfer card, as well as Media Vision's Pro AudioStudio card. The cards are upward-compatible. This means that an 8-bit (sample size) Sound Blaster card will not be able to play 16-bit sound files, but the 16-bit Sound Blaster card can play both 8- and 16-bit audio files. Some cards and systems also do not support the 44.1 KHz sample rate. The function of the sound cards or the built-in sound hardware is to convert the digital sound file into an analog signal. This signal is then made available through output jacks where it can be fed to an amplifier, to powered speakers, or headphones. Although personal computers have a built-in speaker, its sound quality is mediocre.

Many audio cards support a proprietary sound file format. However, the growth of interest in multimedia, and the proliferation of sound cards on the market, have made it necessary to have a standard format so that sound files supplied with programs can play through most systems. For DOS systems, the Sound Blaster format (SND files) has become a standard supported by many. The **WAV** file format is standard for Windows systems.

Note that the data rate for CD-quality audio is higher than the transfer rate of many CD-ROM drives, so although you could store such a waveform file on a CD-ROM disc, you would have to copy it to a hard disk in order to play it. Although multimedia programs and sound card drivers provide some **buffering,** the file data must be delivered to the sound hardware continuously at the data rate specified. At high data rates this places significant demands on the personal computer, especially if some other action is taking place at the same time, such as an animation. Slow computers might not be able to keep up with the high data rates. Only experimentation with your particular application and sound files will reveal whether or not your computer is up to the task.

Most multimedia authoring programs come with a small collection of WAV files that can be used with the program, or with other programs. Libraries of WAV files are also available for purchase.

Computers with sound playback capabilities also have recording capability. Jacks are provided for connection of a microphone or preamp-level input signal (such as one originating from a CD-ROM drive or a tape player). Programs are available that allow you to record sound using the hardware. Windows provides a simple Sound Recorder program as part of the operating system, but there are more sophisticated programs, such as Wave for Windows from Turtle Beach Systems. The quality of your recordings will be affected by a number of factors, but it is crucial that your computer be fast enough to keep up with the data rate (see Table 4) of your recording.

WORKING WITH DIGITAL AUDIO

We will use Turtle Beach's Wave for Windows to illustrate working with digital audio. This tool allows you to edit audio files about as easily as a word processor lets you edit text files.

Figure 24 shows the window the program displays when you choose to record an audio file. In order to record such a file, your computer must be equipped with a Windows-compatible sound card (such as one of those mentioned on page 47), and a microphone must be connected to the recording jack of the card. The controls in the window let you easily select the sample size (called Resolution in Figure 24), the sample rate, and either mono or stereo. The window displays how much disk space you have available for recording, calculated based on the settings of the controls.

To record, you use the same process you use to record with a tape recorder. When ready, press the Record button, and monitor the signal level to make sure it is adequate. When the recording is complete, press the Stop button. It is then possible, using the tape recorder–like controls, to play back the recording or record over it.

Once the file is recorded it will be displayed in its own window, as shown in Figure 25. You can view the file on any time scale. The numbers along the bottom of the picture represent time in milliseconds.

Suppose you would like to delete a portion of the sound file. You do this by selecting the desired portion of the file on the screen with the mouse, as shown in Figure 26, and selecting the Cut command on the Edit menu.

The resultant recording is shown in Figure 27. You can also paste the deleted portion anywhere else in the recording, choosing either to replace what is recorded at that particular location or to insert the portion at that location, moving the rest of the recording forward to make room for the inserted portion.

Although being able to cut and paste portions of recordings in this way is very powerful, the program allows you to do a lot more with the recording. The different effects are shown in Figure 28. Each effect might have a number of options, and you can apply the effect to the whole recording or to part of it. The Gain Adjust

FIGURE 24   ▌ *Wave for Windows Recording Screen*

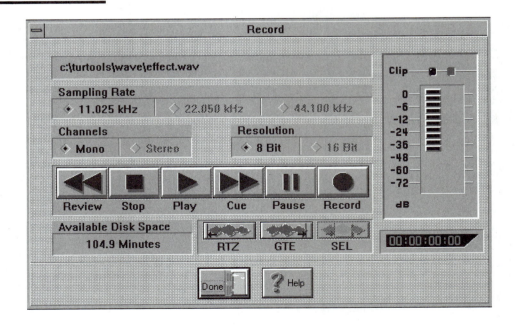

FIGURE 25   *The Recording as Shown in Wave for Windows*

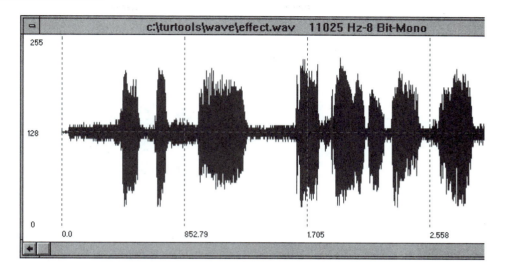

FIGURE 26   *Selecting a Portion of the Recording for Deletion*

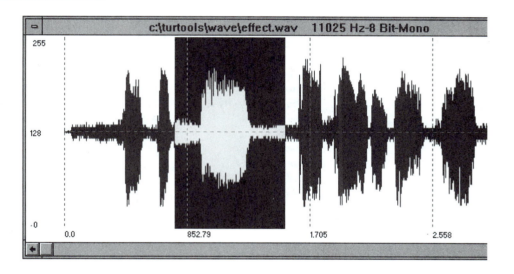

tool, for example, lets you select the amount of start gain and end gain (in case you want the gain, i.e., the change in sound volume, to vary), and whether you want the gain variation to be linear or exponential. The Reverb tool lets you select from more than 25 reverb effects or create your own.

MIDI

Musical instrument digital interface, or MIDI, is a musical note communications specification originally written by musicians to enable different musical instruments to be interconnected. The MIDI **protocol** is known as a *high-level protocol,* meaning that it takes very few bytes to convey the sound information. Whereas a waveform audio file contains a numeric representation of a sound

FIGURE 27   *Recording After the Selected Portion Has Been Deleted*

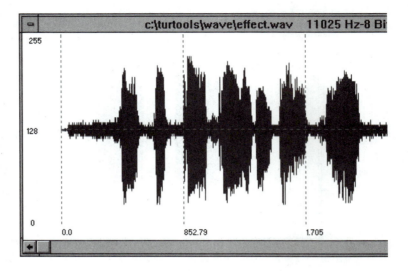

FIGURE 28   *Functions That Wave for Windows Can Apply to a Recording*

Function	Shortcut
Fade In	
Fade Out	
Gain Adjust ...	G
Mute ...	
Equalize ...	E
Frequency Analysis ...	F
Mi**x** ...	M
Crossfade ...	X
Reverse ...	
Invert ...	
DC Offset ...	D
Time Compress/Expand ...	C
Auto Stutter ...	
Di**s**tort ...	
Fla**n**ge ...	
Digital De**l**ay ...	
Rever**b** ...	
Speed **U**p / Slow Down ...	
Normalize	N

waveform, a MIDI file contains coded instructions, such as "play middle C on the piccolo." In order to hear the note, the command must be received by a MIDI **synthesizer,** which interprets the command and generates the sound.

MIDI supports the concept of multiple channels. Each channel can be assigned an instrument sound. There are 16 channels, so MIDI can create the sound of multiple instruments playing simultaneously, as in a band or orchestra. There are 128 defined instrument sounds, such as piano, piccolo, bassoon, pipe organ, violin, viola, trumpet, and so on. There are also 47 defined percussion instrument sounds (drums, etc.), which are assigned to channel 10. Although each channel can only be assigned a single instrument at any one time, it can play multiple notes of the instrument simultaneously, in the same way that a pianist may play several keys at the same time. MIDI also allows you to individually control the volume, the pan (position of the instrument in the stereo field), the reverberation, and so on of each channel.

A MIDI synthesizer uses a technique called *FM synthesis* to generate sound. A measure of the quality of the synthesized sound is the number of operators (frequency generators) employed. Thus a two-operator FM synthesizer will not sound as good as a four-operator FM synthesizer. FM synthesis does not do a good job of generating certain instrument sounds, such as piano. To overcome this, some synthesizers use a technique called *wavetable synthesis.* This technique consists of recording the actual sound of the instrument, digitizing it (see "Sampling" on page 8), and storing it in the synthesizer's read-only memory (ROM). When a MIDI command calls for a note to be played with that instrument's sound, the synthesizer will use the stored sound as the basis for creating the note. In other words, more-sophisticated MIDI synthesizers employ waveform audio techniques to generate the sound. Fortunately, it is not necessary to store every note the instrument can play. The synthesizer transforms and shapes the sound waveform to create other notes, tempos, attack velocities, and so on.

MIDI files are very compact; a typical MIDI data rate is 12 kilobytes per *minute* of music. Another feature of MIDI is its flexibility. For example, an editing program can quickly change the tempo of the music, or which instruments are assigned to which channels at a specific point in time. You can easily mute one or more channels, or change the volume of all or of any specific channel. This editing capability is simply not available with waveform audio. You can also take an existing MIDI file and add instrument sounds on vacant channels, or compose music by composing one channel at a time.

How can MIDI files be so much more compact than waveform audio files? What is the catch? Think of MIDI as a form of compression. A waveform file is "dumb." All it takes to play it back is some hardware that can "connect the dots" (called a *digital-to-analog converter,* or DAC). A MIDI file, on the other hand, might have a command such as "assign a grand piano to channel 2" followed by "play a C-sharp on channel 2." The hardware receiving the command must be sophisticated enough to know how to generate each note, produce a grand piano sound and many other instrument sounds, place the instrument in its correct position in the stereo field, generate reverberation effects, and so on. We can stretch this concept of compression even further: Imagine a musical file that contained the single command "Play Beethoven's Fifth Symphony." This single command file would be very small, but the hardware required to play it back would be very complex, enough to be impractical, as it would have to know how to play any composition you might ask of it.

Once an instrument is assigned to a channel, it only takes 6 bytes to play a note. Separate 3-byte commands turn each note on and off. The delay between the two commands determines the duration of the note. Each command specifies the channel, the note, and the note's velocity (how hard the note is played).

MIDI files can be generated by hand or by computer programs. However, the easiest way to generate a MIDI file is to record it using a MIDI instrument, such as a MIDI keyboard, horn, flute, drum, guitar, violin, and so on. These MIDI instruments contain electronics that convert the musician's actions into a stream of MIDI data on the MIDI interface, which the computer can record and store.

There is more than one MIDI file format; the two most prevalent (for computer multimedia) are SMF (Standard MIDI File) type 0 and SMF type1. Most programs can handle either type. Another specification, General MIDI Mode, specifies the

FIGURE 29   MIDI Tune-Up Program

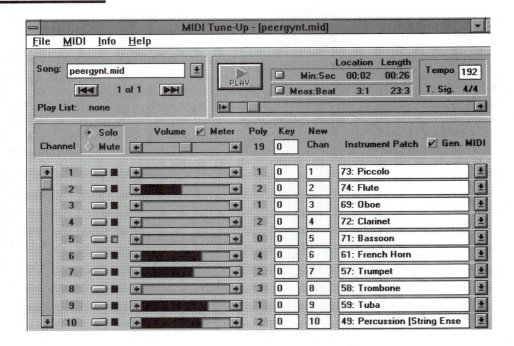

contents of MIDI files and assigns standard numbers for instrument sounds, as well as specifying channel 10 for percussion. This prevents the problem of the file playing with a violin sound in one system and with a piano sound on a different one. The MIDI specification is controlled by the MIDI Manufacturer's Association (MMA) and the Japan MIDI Standards Committee (JMSC).

Most audio cards contain built-in MIDI synthesizers, so they can play MIDI files as well as waveform files. The synthesizers vary in the number of channels they support and the number of instrument sounds available. The sound signal generated by the synthesizer is provided at the same output jacks as the sound signal generated by the waveform circuitry. The sound generated by the built-in synthesizers is good, but no match for the sound generated by expensive stand-alone synthesizers. If you require high-quality sound, you can purchase a MIDI card and a driver that enables the computer to generate the signals an external MIDI synthesizer needs in order to generate sound. Many audio cards also provide a MIDI jack, thus allowing you to play through the card's own synthesizer or through an external synthesizer connected to the card. With the proper software, you will then be able to play and record MIDI files.

A more recent implementation uses a special software driver that sends the MIDI commands to an external synthesizer through the computer's **serial port.** This approach has the advantage of not requiring a special card to be installed in the computer, thus making it suitable for adding MIDI capability to laptop computers.

A limitation of MIDI is that it can only reproduce the sounds that are programmed into the synthesizer. It cannot reproduce the human voice or the sounds of animals. In addition, the quality of the sound is related to the quality (i.e., the cost) of the synthesizer.

Due to its small file size and good sound quality, MIDI has become a popular sound adjunct to presentation programs. MIDI song files are generally supplied with multimedia authoring programs. In addition, companies such as Passport Designs sell libraries of MIDI music files.

WORKING WITH MIDI

You can use a program, in conjunction with a MIDI instrument, to create a MIDI composition. The software will enable you to edit the composition and to make whatever transformations you might want. However, MIDI is not limited to musicians. As long as you have an audio card with a MIDI synthesizer on it (most sound cards do), you can play MIDI files. You may also avail yourself of a number of MIDI programs that do not require knowledge of music in order to work with MIDI compositions.

Figure 29 shows one program, MIDI Tune-Up, written by FM Software and distributed by Turtle Beach Systems. Once a MIDI file is loaded, the program displays the instrument assigned to each channel. While the song is playing, bar meters show the volume for each channel. The Poly column shows polyphony, or the maximum number of notes the channel is asked to play simultaneously. Notice that for the music playing in the picture *(Peer Gynt)* there will be a point at which the trombone, on channel 8, will be required to play three notes simultaneously. The program allows you to solo any channel, to mute one or more channels, to change the tempo of the song, and to control the volume of the song as a whole and of each individual channel. You can change the instrument assigned to any channel (Figure 30), as well as the effects for each channel (Figure 31). In Figure 31, Pan indicates the position of the instrument in the stereo field.

There are even programs that will generate MIDI music for you. One example is Soundtrack Express, which is pictured in Figure 32. This program has you specify the style, personality, instrumentation, duration, tempo, key, and so on for a song, and then it generates one for you. If you do not like the song, you can ask it to generate a different one from the same parameters, or you can modify the parameters and try again. The program creates files in the general MIDI format, and the computer is not particular about copyright, so the songs you generate can be distributed freely with no royalty obligations.

FIGURE 30 ▌ *Changing the Instrument Assigned to a Channel*

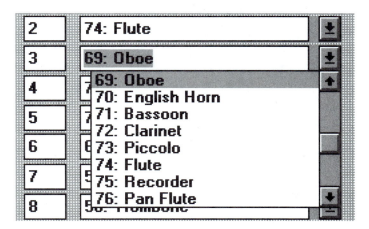

FIGURE 31   *Changing the Effects Assigned to a Channel*

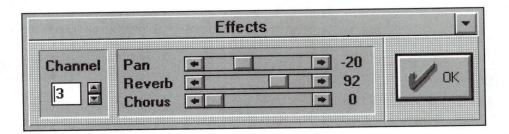

FIGURE 32   *Soundtrack Express Screen*

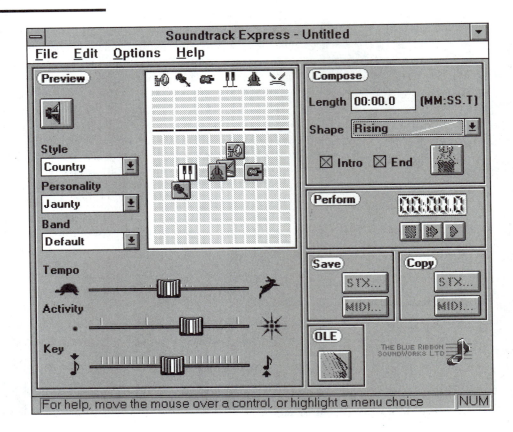

7

Analog Video

It may seem odd to talk about analog video in the middle of a discussion of digital multimedia. However, analog video has played a significant role in the application of multimedia to education.

Although a videodisc (also called a laserdisc) looks like an overgrown compact disc, and both are read by shining a laser onto their surfaces, they are fundamentally different. The information on a videodisc is recorded in analog format, just as is the audio information on a vinyl record. Because the information is in analog form, and computers deal only with digital data, a special hardware card is needed in the computer if the video is to be displayed on the computer monitor. The video cannot be edited; except for being able to specify where playing starts and stops, you are not able to modify the video on a videodisc.

The use of videotaped materials has been extensive in higher education. However, applications combining the video cassette recorder (VCR) with the computer have been rare, both because of the expense and rarity of VCRs capable of computer control, and because the videotape does not lend itself well to hypermedia applications, given the difficulty in quickly accessing different frames on the tape. Suffice it to say that a few models of computer-controlled VCRs, known as PC VCRs, do exist, and that the **MCI** specification (see "MCI" on page 210) does define commands for such devices.

The rest of this chapter will be dedicated to the discussion of videodiscs and videodisc players. Videodiscs initially did not capture the fancy of the consumer. However, they have enjoyed wide acceptance in higher education, and have also been used extensively by business for training. Lately the increased availability of high-quality TV sets, coupled with a growing demand for high video quality, have caused a resurgence of demand for videodiscs in the consumer market.

FIGURE 33   ▌   *Videodisc*

The Videodisc

The videodisc (Figure 33) has a diameter of 12 inches, compared to the CD-ROM disc's 4.7-inch diameter. It is recorded using similar technology. A videodisc contains a video track along with high-quality audio tracks. The two forms of videodiscs, CAV discs and CLV discs, will be described below.

Videodiscs have several advantages over videotapes:

- **Video quality.** The quality of the video provided by videodiscs is higher because they are typically recorded with more lines of resolution than a videotape is.
- **Durability.** Videodiscs are not affected by magnetic fields nor do they deteriorate with repeated use.
- **Random access.** The movable laser head in the player can quickly seek to any location on the disc, thus making the videodisc well suited for use in hypermedia applications.

CAV DISCS

CAV (constant angular velocity) discs are the ones found most often in an educational setting. Each side of the videodisc can contain 30 minutes of video, along with 30 minutes of two-track audio. Because the video is composed of individual still frames, which play at the rate of 30 frames per second, simple math tells us that one side of a videodisc contains 54,000 frames. Many CAV videodiscs contain combinations of images intended to be viewed as stills, as well as sections intended to be viewed as full-motion video. A distinguishing characteristic of CAV laser videodiscs is that its tracks are laid out as concentric circles and that each frame takes up exactly one track, so it takes one revolution of the disc to read a frame. This means it is very easy to "freeze" the display on one frame. The player simply reads the same track over and over again, and thus 30 times per second the same frame is placed on the screen.

The two audio tracks on a CAV disc are independent. They can provide stereo sound, or each track can present the material in a different language, with the student able to select which track to listen to. Another approach is to have each track contain material with different levels of difficulty.

CLV DISCS

CLV (constant linear velocity) discs are most often used to play back movies, symphonies, and operas. They are more prevalent in the entertainment and consumer marketplace. About twice as much video can be recorded onto a CLV disc, but it is not as frame-oriented as the CAV disc. The less expensive players are not able to seek to a specific frame, and they may not be able to "freeze" a single frame. The only way to "freeze" a frame on a CLV disc is to have memory in the player where the frame can be stored.

On a CLV disc the information is recorded in one continuous spiral track, just as on a CD-ROM disc. Most players can play both CAV and CLV discs.

Videodisc Player Levels

There are several types of videodisc players, which differ in their interactive capabilities. These are known as Level 1, Level 2, and Level 3 players.

Level 1 players are similar to home video cassette recorders (VCRs). Using front panel controls, or a remote control, you can play a disc, seek to a different portion of the disc, pause the player, change the playback speed, and so on. Level 1 players are mostly intended for home use. The addition of a barcode reader to the players is a newer feature that makes it possible for a professor, using a printed guide, to play specific portions of the videodisc without having to manually seek to the desired passages.

Level 2 players incorporate circuitry that controls the player in response to commands recorded on the videodisc itself. For example, you might choose from a menu and, in response, the player will play a specific sequence on the disc that is encoded on the disc itself. Level 2 players have never achieved much popularity, probably due to the lack of a sufficient number of encoded videodiscs.

Level 3 players contain circuitry and an interface so that they can be connected to and controlled by a computer. This is the type of player prevalent in higher education.

The Videodisc Player and the Computer

Early applications of videodiscs employed the computer to control the player, but the video would be displayed on a TV monitor separate from the computer monitor, as shown in Figure 34. A newer approach is to have the video brought into the computer, where it is displayed in a window of the computer's display. This is sometimes called *overlay video,* because the video overlays a portion of the computer screen. This setup is shown in Figure 35. This approach has a number of advantages, among them the need for one less screen. Other advantages include the capability to scale the video, to display only portions of the video image, and to add computer graphics elements to the video. For example, a program might draw an arrow over the video pointing to an item of interest. It can also zoom in on the video, concentrating on a specific portion of the image.

FIGURE 34 | *Setup for a Videodisc Player with Separate Video Monitor*

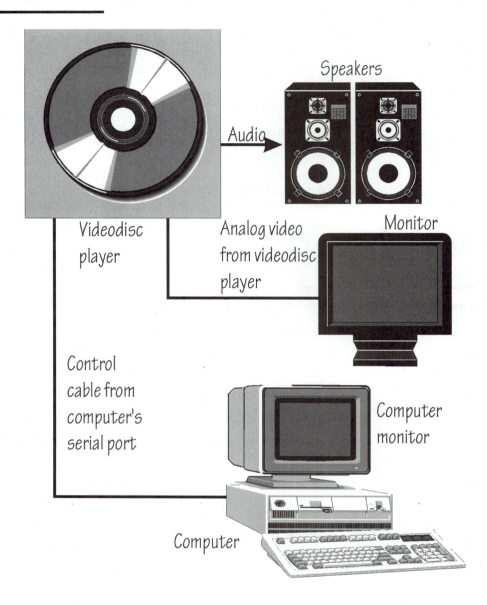

Speakers

Audio

Videodisc
player

Analog video
from videodisc
player

Monitor

Control
cable from
computer's
serial port

Computer
monitor

Computer

The video is brought into the computer through a special card that digitizes it, scales it, and mixes it with the computer's own video signal. The composite image is then routed to the display. The card does all the processing, so there is very little extra demand placed on the computer's processor or memory. These cards also provide the ability to capture the current frame and write it to a disk file. The file might be read at a later time, displaying the frame on the screen, even if the videodisc player is no longer connected to the computer.

There are a number of these video cards on the market, such as Creative Labs' Video Blaster, and cards from VideoLogic.

The computer controls the videodisc player through a connection between the player and the computer's serial port. A driver program is needed that translates your program instructions into commands the videodisc player can understand. Some commands the player can execute include play, pause, stop, play back-

FIGURE 35 *Setup for Overlay Video*

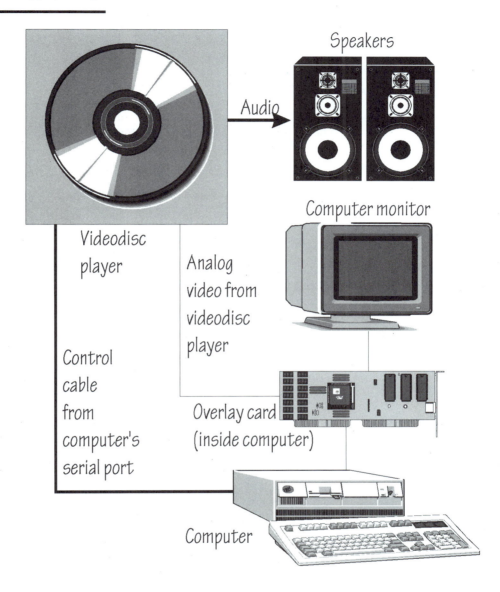

wards, seek, play fast or slow, eject the disc, provide the current play position, and turn the left and right audio channels on or off. Controlling a videodisc player is simple and does not consume much of the computer's time.

The audio output of the videodisc player may be connected to an amplifier, powered speakers, or appropriate headphones. The output may also be connected to an audio card in the computer for mixing with other audio signals.

The majority of videodisc players sold in the United States are made by Pioneer and Sony. There are a number of models in each manufacturer's line. To hook up the player to your computer, in addition to the electrical cabling, you will need two driver programs. One driver controls the videodisc player (the newer drivers can control any model player in a manufacturer's product line, but a Sony driver will not work with a Pioneer player, and vice versa). The other driver controls the operation of the video mixing card. This driver should come with the card itself. Note also that there are separate drivers for DOS and for Windows.

One final note of caution: Different players use different communication speeds (bits-per-second rates) on the serial port connection, so you might have to configure the driver for the player's set speed. The most common speeds are 4,800 bps and 9,600 bps.

Creating Your Own Videodisc

Many educators author their own videodiscs, for the same reasons they would author a CD-ROM disc (page 43). Imagine how much less space a single videodisc occupies compared with 54,000 slides! The random-access nature of the videodisc also makes it easy to write applications based on the contents of the disc.

Videodiscs are created from videotapes, so whether you are compiling motion sequences or stills, your source material must be recorded on a 1-inch videotape before disc mastering. This master videotape is used to create a checkdisc, which is a "test" version of the final master videodisc.

Checkdiscs are made of plastic or glass and are thinner and less durable than a regular videodisc. Most facilities can create a checkdisc in 24 hours once your tape reaches them, for a cost of around $300. If you only need one or two copies of the disc, it is cheaper to make a couple of checkdiscs than to create a master. However, the image quality of the checkdisc will be lower than that of a disc produced from a master.

After reviewing the checkdisc, you can make changes to your 1-inch master videotape. If your master videotape is changed, it is a safe practice to cut another checkdisc before pressing the master videodisc. When you are satisfied with your checkdisc, a master disc will be made. Mastering facilities are operated by companies such as Pioneer, 3M, Sony, PDO-Phillips, or DuPont Optical. The master disc is made of thick glass and is used to press the videodisc replications, in a manner similar to a molding machine. The mastering/replication process typically requires 10 business days. You can anticipate a cost of about $2,000 for a one-sided master or $3,600 for a two-sided master. Replications cost between $10 and $20 each, depending on the number of copies you order.

As was the case with CD-ROM discs, you can purchase a machine to make videodiscs in-house. Sony, for example, has a videodisc recorder (LVR-3000N). Again, such a purchase may be justified on a departmental or campus basis if there is significant videodisc-authoring activity taking place.

SOME USEFUL HINTS

In some ways creating a videodisc is not as tricky as creating a CD-ROM disc. There are no performance concerns to worry about, as the video signal is always delivered by the player at 30 frames per second. If you have sets of related stills, it is obviously advantageous to place these close together on the disc to minimize access time.

Keep in mind the "weakest link" rule. The quality of the image and sound of your videodisc is not going to be better than the quality of the video and sound you used to create it. The statement that the quality of video from a disc is superior to that from a tape (page 56) refers to the **"bandwidth"** of the medium. A professional recording, such as for a theatrical movie, will look better on disc than on tape because in both cases a very high quality signal was used, and the disc can preserve more of the signal than the tape can. Do not expect top-quality video if you take your video with an inexpensive camcorder, for example. Also, consider checking such items as color balance and audio levels if you are splicing together material from many sources.

Don't be confused by the labeling of videodiscs. Labels are often placed so you can read them when the disc is in the player, that is, just the opposite of record albums. Whereas record albums are played with the needle on top of the exposed side, a videodisc is read from the underside. On a one-sided videodisc, for example, the label is placed on the side the laser doesn't read. On a two-sided disc, the information is stored on the side opposite the corresponding label.

8

Digital Video

Perhaps no multimedia technology offers so much potential as digital video, yet because the technology is still in its infancy, the potential is unrealized. We define digital video as video that is stored in the computer in digital format. Thus the video is contained in a computer file, no different (except for content and size) than all other computer data files. It can be stored on the computer's hard disk, on a CD-ROM disc, or it can arrive over a network to which the computer is connected.

The Digital Video File

The file contains all of the frames required to make up the movie. It also contains, in most cases, one or more sound tracks, again in digital format. When a movie is played, the audio information is separated, and sound is generated by separate audio circuits, most often the same audio circuitry that processes WAV (waveform) files (page 46).

Information in the digital video movie file is almost always in compressed form. As we have seen (page 14), uncompressed video takes up enormous amounts of space and is not practical for use in personal computers. A number of different compression techniques are used, some proprietary, some open (e.g., MPEG). Movies are stored in a number of color depths, such as 8-bit color and 24-bit color. The minimum acceptable color depth is 8 bits, or 256 colors.

Digital video is very attractive for a number of reasons. Because the information is in digital form, it is editable by the computer. A number of software programs are available (such as Adobe Premiere, D/Vision and D/Vision Pro) that allow you to easily assemble video movies by splicing together segments from different files, cutting unwanted segments, mixing video tracks, mixing audio tracks, overlaying computer-generated graphics and text on the movie, creating transition effects between scenes, and so on. Just about all of this capability is available for analog video (videotape), but the cost of the equipment to perform such editing is well beyond the financial means of most of us.

Another advantage of digital video is that it can be sent over computer networks, can be stored on a server, and can be copied with no loss in quality, because digital video is in the form of a computer data file and can thus be handled as such.

Hardware Motion Video

Hardware motion video movies rely on a hardware card in the computer to handle the decompressing of the movie as it is played. The use of the card frees the computer's processor from doing this task, which results in the ability to deliver higher-resolution video and a higher number of frames per second. The disadvantage of such cards is their cost and the requirement that every playback machine have one.

ACTIONMEDIA II

The ActionMedia II card, jointly developed by IBM and Intel, contains a specialized processor, the Intel i750, which is designed specifically to handle the video compression and decompression tasks. The card also contains a **digital signal processor (DSP)** dedicated to compressing and decompressing the audio signal using ADPCM compression (page 22). The technology employed by ActionMedia II is known as **digital video interactive,** or **DVI.**

The ActionMedia II card can deliver 30-frame-per-second (full-motion) video. The video resolution is 256 × 240 × 16.7 million colors, which can be scaled to any size, including full screen. The ActionMedia II card also handles the mixing, or overlaying, of the movie video with the computer-generated screen, so that the motion video can be played in a window on the computer screen. Stereo audio is supported. In addition to full-motion video, the ActionMedia card is also capable of decompressing still images using JPEG algorithms. Drivers are provided with the card for both OS/2 and Windows systems. Two types of video are supported by this card, PLV and RTV.

PLV   Production-level video (PLV) is video that is compressed using a supercomputer at Intel's facilities. To create such a video you submit your master tape to Intel and receive back the PLV file. The cost of this service is high, so it is only used for commercial purposes. On playback, PLV quality is similar to that provided by an average VCR. Compression ratios can be as high as 200 to 1. PLV files require about 150 kilobytes per second of video, so it is possible to record them onto CD-ROM discs for distribution. The McGraw Hill Encyclopedia of Mammalian Biology utilizes PLV stored on a CD-ROM disc.

RTV   Real-time video (RTV) is video that is compressed in real time by the ActionMedia II card. To create an RTV file, you need to obtain the ActionMedia II capture card, which plugs into the ActionMedia II card. This card digitizes incoming audio and video, which is then compressed by the processors on the ActionMedia II card and stored on the computer's hard disk.

RTV can approach the quality of PLV, but the data rate (or data storage requirement) is two to three times as high. This is understandable considering that the processor on the ActionMedia II card has only 1/30 of a second to compress each frame, so it does not have sufficient time to optimize compression. If lower data rates are required, they can be achieved at the expense of some loss in quality.

The advantage of RTV is that you can capture and digitize your own video, without resorting to expensive off-site processing. The disadvantage is that the data rates preclude the inclusion of RTV on a CD-ROM disc, unless you can require the use of a multispeed CD-ROM drive, or if the user can copy the digital video file off the CD-ROM onto a hard disk.

MPEG

The emerging standard for hardware motion video is MPEG, the standard developed by the Motion Picture Experts Group. There are two MPEG standards, one aimed at providing motion video for personal computers and CD-ROM players attached to television sets, the second aimed at broadcast-quality television. MPEG I, the one aimed at personal computers, plays 320 × 240 resolution video at 30 frames per second. MPEG boards, such as Sigma Designs' Reel Magic, can scale the video to full screen (640 × 480) by a process of interpolation. A CD-ROM can hold 72 minutes of MPEG-compressed video, and several motion pictures are available on CD-ROM. Currently MPEG compression hardware is still expensive (although prices are dropping), so it is not yet practical for consumers to create their own MPEG videos. The consumer-oriented boards are playback capable only.

MPEG employs a number of compression techniques to achieve its 200-to-1 compression ratio. First, the image resolution is reduced to 320 × 240, which reduces the amount of pixels in each frame. The signal is represented in a luminance (brightness) and chrominance (color) format. Since the eye is more sensitive to brightness than to color, some color information can safely be discarded. The discrete cosine transform is performed on the frame, followed by the discarding of the higher-frequency components of the transform, followed by quantization. Finally, the resultant information is compressed again using Huffman encoding.

MPEG also performs interframe compression by only saving the changes between frames, not the frames themselves. Every few frames, a key or reference frame, in other words, a frame that does not depend on the information from previous frames, is saved. MPEG also uses a motion-estimation algorithm to aid compression. Think of a video of a car moving down a street. Most of the background does not change from frame to frame, so interframe compression eliminates the redundant information. Now let's consider the car itself. The position of the car will change from frame to frame because of its motion. However, the image of the car will remain relatively static (the shape of the car does not change). Motion estimation detects this, saving only the amount by which the block of pixels has moved, not the pixel block itself.

MPEG can also be decompressed using software. However, the process is so complex that today's personal computers do not have enough processor power to decompress 30 frames per second at full resolution. The MPEG standard is much more complex than this brief discussion might imply. MPEG allows for different resolutions, aspect ratios, and frame rates. MPEG 1 and MPEG 2 are specific implementations of the MPEG standard.

Software Motion Video

Software motion video refers to the playback of digital video files without the aid of any specialized hardware (for the video portion of the file). The computer's own processor handles all the tasks associated with retrieving the file, decompressing the video information, and painting that information on the screen. Any audio tracks present in the movie file are sent to an audio card. If no audio card or hardware is present, the movie plays without sound. Examples of software motion video include Apple's QuickTime, Microsoft's Video for Windows, and IBM's Photomotion, which runs on DOS systems. In order to play a video file, the computer must be able to perform the following tasks:

- Read chunks of data from the file and place the data in memory. This process takes place every time the buffer in memory (the number of bytes left from the last-read operation) gets low, which is quite frequently.
- Separate the video data from the audio data, and transfer the audio data to the audio card (if there is one).
- Process the video data through the use of fairly sophisticated and computationally intensive decompression algorithms. As the data is decompressed, the results are stored in another area of memory.
- Map the video from the stored frame size to the size of the window the video is to play in. Given the state of the technology, it is strongly recommended that movies be played at their default frame size. Trying to play a movie in a larger window will slow down the frame rate dramatically, and the quality of the picture will deteriorate.
- Move the full frame of video to the video buffer (where the picture currently on screen is stored) for display.

Not only does this processing burden overwhelm today's processors, but the volume of data overwhelms the computer's internal data bus. For this reason, full-screen, full-motion video is not yet achievable with software motion video, even on fast computers. Video is limited to small window sizes (generally 160 × 120, but sometimes up to 320 × 240), shallow color depth (256 colors), and a frame rate of 10 to 15 frames per second. Sound is generally 8-bit monaural.

Yet with all these compromises, file sizes are still quite large. A 5.3-second Video for Windows clip that plays at 15 frames per second, with 256 colors, in a 160 × 120 window (1/16 of a **VGA** screen), accompanied by monaural 8-bit, 22 KHz sound, takes almost 1.1 megabytes of disk space.

Software motion video is scalable, which means that the playback rate of the video is adjusted to the capabilities of the computer. If a video file is stored as 15 frames per second but the playback computer cannot maintain this rate, playback might take place at 8 or 10 frames per second, with the playback software discarding some of the frames so that the movie will still play in the same amount of time, which is crucial for maintaining synchronization with the sound. The limitations of operating systems such as Windows also cause synchronization to be poor at times.

The key to software motion video playback, then, is a very fast computer. If software motion video is a requirement, purchase the fastest machine you can afford.

Software motion video will only get better, though, as better algorithms are implemented and personal computer performance continues to improve. It is reasonable to expect that sometime in the future (but not within the next few years), personal computers will be able to play full-motion, full-screen, 16.7-million-color video without problems. Until then, those attracted to the advantages of digital video can achieve better results by investing in hardware motion video playback cards.

Video Capture

Although you can purchase libraries of video clips in various formats (for both hardware and software motion video), most people want to record their own movies. A zoologist wants to record the behavior of animals to illustrate his or her lectures. A psychology professor wants to record people's reactions to experiments, as well as behavior traits that will be used in computer-based instructional material. A chemistry professor wants to record chemical reactions for incorporation into instructional software. A current events teacher wants to record news clips to enrich classroom presentations.

Specialized hardware is always required to enable you to create movies. The hardware consists of a capture card that plugs into the computer; the card receives the analog signal (**NTSC, RGB, PAL,** etc.) from a broadcast, a VCR, or a camcorder, and digitizes it. Some cards also provide **S-VHS** input; S-video is a higher-quality video signal, so its use will result in a better video. The hardware passes the digital data to the processor, which stores it in a file. Hardware motion video products either include capture facilities or have available a capture card, which plugs into the playback card.

Although this is a new technology, there are already more than a dozen cards available for personal computers to capture video for Video for Windows.

The Windows cards use the Microsoft Video for Windows video capture (VidCap) program to capture the video. This program allows you to select color depth (all cards are capable of 24-bit color), frame rate, and video dimension, as well as to select the audio capture specifications—the audio is captured using a separate audio card that is not a part of any of these products. Although the audio is captured by a separate card, the video recorder software combines the digitized video and audio into a single disk file.

Capture of 8-bit, 160 × 120 video at 15 frames per second, with audio, should not be a problem on a fast machine. Capture of 16-bit color also works fairly well. Capture of larger video frames, such as 320 × 160, is much more problematic. If the computer cannot keep up with the video, the cards respond by dropping video frames.

There are two ways to capture video. The first is real-time capture, where the video is coming from a video camera, live broadcast, or VCR. The other way is to capture a single frame at a time, allowing the computer as much time as it needs to process and compress the frame. This type of capture is only feasible with a frame-accurate video source (one where you can reliably play back any specific frame) such as a videodisc player. Frame-by-frame capture is feasible mostly for commercial title developers, who can afford the extra step of pressing a videodisc from the source videotape.

COMPRESSION

There are two ways to compress video. The first is called *symmetric,* meaning that it takes the same time to compress as to decompress the video. Since playback is done in real time, **symmetric compression** implies that the video is compressed in real time, as it is received. This also implies that the compression is performed by hardware, as processors do not yet have the speed to perform good compression in real time. With **asymmetric compression,** the compression step takes longer than the decompression step, so compression is not performed in real time.

Both QuickTime and Video for Windows provide their own compression algorithm; they also allow vendors to install proprietary compression modules into the product so that you can choose, from the Video Editor program, which module to use. On playback, the Video for Windows and QuickTime software automatically select the proper decompression module. The set of compression and decompression modules is known as a **CODEC.**

Some products provide symmetric (real-time) compression. Some provide software compression, meaning that the processor in the personal computer performs the compression of the raw frames received from the capture card.

The Video Spigot card, from Creative Labs, uses this approach. Other cards provide hardware compression on the card itself, so that the computer processor's only video task is to save the compressed frames. The amount of data traffic inside the computer is lower for the same reason. These cards can therefore achieve higher frame rates or larger frame sizes without overloading the processor. The Intel Smart Video Recorder (which employs technology based on DVI) is one such card. On a fast computer it can record 24-bit color with no loss of frames.

SCALABILITY

As previously mentioned, scalability is the ability of the playback software in software motion video to adjust the frame rate to the capabilities of the computer. The same movie might play back at 15 frames per second on one computer, and at 8 frames per second on another. Scalability is achieved by discarding frames during playback.

Note that it is not possible to scale upwards. A video captured at 15 frames per second will play no faster than 15 frames per second, no matter how fast the processor. If a video capture card has to discard frames during recording, those frames are lost forever, no matter how fast the playback machine might be. The result is very jerky playback. The lesson here is to use the fastest computer you can get your hands on for video recording.

INDEO

The most popular software motion video technology for the Windows platform is Intel's Indeo, a video data format derived from the DVI technology. The Intel Smart Video Recorder card contains the circuitry to digitize and compress the video signal in real time. Decompression and playback is handled by software, with no need for specialized hardware. Because the Smart Video Recorder has the compression hardware on the card, it performs symmetric compression. Other boards perform some compression on the card (such as reducing the image resolution, and skipping frames), relying on the computer's processor to perform additional compression.

SINGLE FRAME VIDEO CAPTURE

The Video for Windows Video Editor program allows you to save a frame from a movie as a separate image file, so you can use it as an illustration in other programs. DIB and PCX formats (see page 27), among others, are supported. Some video capture cards are also capable of capturing single frames directly, meaning that they can, upon command, capture the current frame in the video signal stream and write it as a DIB or PCX file.

Advantages of Digital Video

Digital video has not yet achieved the quality level of analog video. Professors do not consider the so-called postage-stamp–sized video (160 × 120) adequate for instructional purposes. However, as computers become faster and the algorithms improve, digital video is improving. It boasts a number of advantages:

1. The video is stored as a standard computer file. Thus it can be copied with no loss in quality, and also can be transmitted over standard computer networks.

2. The video information can be easily edited. You can cut and paste portions of video as easily as you cut and paste words in a word processor. You can overlay videos, add titles, mix in computer graphics, filter the video, correct its color, contrast, and brightness, and add professional transition effects between video clips. Such editing capability for analog video is well beyond the budget of faculty and consumers.
3. Software motion video does not require specialized hardware for playback.
4. Unlike analog video, digital video requires neither a video board in the computer nor an external device (which adds extra cost and complexity) such as a videodisc player.

Working with Digital Video

In order to illustrate the power of digital video editing, we will use Adobe Premiere to put together a digital video. We start out with two clips, one of the space shuttle launch and another of a view of earth from space. We also have a bitmap, created in Windows Paintbrush, consisting of a blue background with white letters that spell out "Fantastic Voyage." We will use this bitmap as a movie title.

Figure 36 shows the Project window in Adobe Premiere, one of several windows in the program. It is into this window that the components of the movie are loaded. The window indicates we have two movies and a bitmap loaded. A thumbnail image of the first frame of each movie is shown. Characteristics of the components are shown, such as movie duration and image sizes. The squiggly line below the thumbnail image of each movie represents the sound track for that movie.

If we double click on the thumbnail image of the Slaunch.avi clip, we open a window where we can specify the starting and ending point of the clip

FIGURE 36   *Adobe Premiere Project Window*

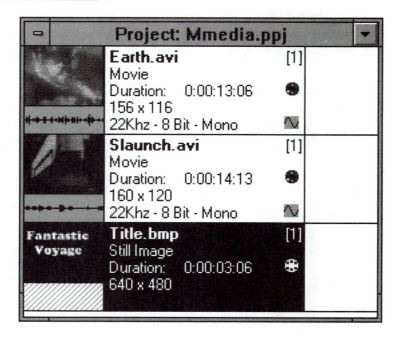

Setting the Starting Point for a Clip

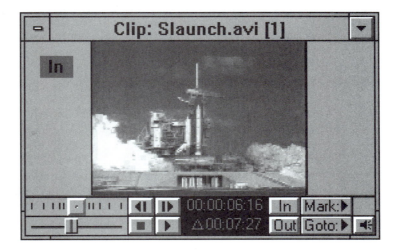

(Figure 37). Here we set the starting point at 6.16 seconds into the clip, where a view of the shuttle from a distance shows. There are several other ways of selecting portions of clips to use in movies.

The next step is to construct our movie, by moving the component elements into the Construction window (Figure 38), which has the form of a light table. You move the components to the Construction window by dragging them from the Project window and dropping them into place. Notice that Premiere provides you with two video tracks (A and B) and an overlay track, where we placed the title. The timing of the movie (in seconds) appears across the top. We have dragged the right edge of the bitmap on the overlay (Super) track toward the right, to increase the amount of time it is shown.

What we would like to do is have the title slowly fade away, letting the clip in track A show through. To further enhance the effect, we want to make the letters in the title transparent, so that the clip in track A shows through. These are sophisticated effects yet easy to achieve. We select the title in the Construction

Construction Window

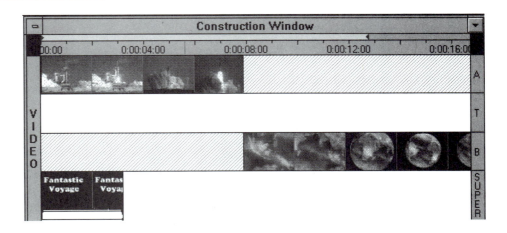

window and choose Transparency from the Clip menu. In the dialog box of Figure 39 we select the white color within the title to be transparent.

To cause the title to fade, we drag the "volume" slider beneath the title (highlighted in Figure 41) so it starts out at the top and ends up at the bottom. This causes the intensity of the title to fade during that time period. By the way, to keep the example simple, we are not dealing with the audio tracks in this example. However, they can be edited just as easily.

Let's finish the movie by adding a transition effect between the two clips on tracks A and B. From the Transitions window (Figure 40) we can choose from several dozen effects. We picked the Spin Out effect (not shown in figure), and placed it twice in the T (transitions) track. After changing some of the transition's parameters (made available by double clicking the transition on the T track), we end up with the Construction window as pictured in Figure 41.

The duration of each transition, like the duration of the title, can be adjusted by dragging either side of the transition box in the T track to make it longer or shorter.

We can see what the movie will look like by pressing the Enter key, which brings up the Preview window and plays a rough sketch of the movie. However, you can get a good idea of what the movie will look like by examining the Construction window (Figure 41). Draw a vertical line anywhere in the window. Everything intersected by the line will play simultaneously.

Figures 42 through 48 show frames from the finished movie, as follows:

1. The movie starts with the title showing, with the clip in track A showing through the transparent letters (Figure 42).
2. As the clip plays, the title fades away (Figure 43).
3. The clip on track A continues to play by itself (Figure 44).
4. Due to the transition effect, the clip on track A spins out (Figure 45).
5. Due to the transition effect, the clip on track B spins into the picture (Figure 46 and Figure 47).
6. The clip on track B continues to play until the end (Figure 48).

The work of creating this movie in Adobe Premiere should take about 5 minutes, once you are familiar with the program. When you are satisfied with your work, you ask Premiere to create the final movie file (Figure 49). This process will take the computer at least several minutes. When your movie is complete, it will be in the form of a Video for Windows movie.

Creating such a movie by conventional means would take equipment costing tens of thousands of dollars. As computers become faster, you will be able to work with movies larger than the 160 × 120 format used in this example.

Getting Started

As you have probably concluded by now, there is a lot to consider as far as digital video is concerned. The following points may help out:

1. If you are only going to use a program that incorporates prerecorded Video for Windows movies, then you should not have to do anything. The program should include the playback modules of Video for Windows and install them for you when the program is installed.
2. Do you need digital video? Or will analog video do?
3. Consider what quality level of video playback you need. If your application requires full-motion video, or video frame size greater than 160 × 120, you will have to work with hardware motion video. Is the cost of the playback card for each machine a roadblock?

FIGURE 39   *Transparency Setting for Title*

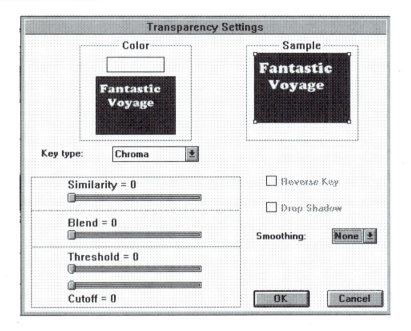

FIGURE 40   *Premiere Transition Effects*

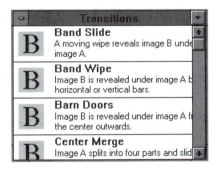

FIGURE 41   *Fading the Title, Transition Effects*

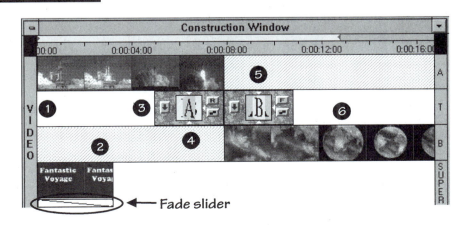

4. If software motion video will do, purchase a video capture card for your computer. If you can afford it, purchase one that includes on-card hardware compression of the video (the video will always play back with software decompression—the card is only needed to capture the video). You must

FIGURE 42 | *First Frame of Movie*

FIGURE 43 | *Title Bitmap Fades*

FIGURE 44 | *Clip on Track A Plays by Itself*

also have a copy of the Video for Windows program, which should be included with the capture card.

5. Consider purchasing a good video editing program (see Chapter 11) so that you can process your movies after you capture them. Popular choices

FIGURE 45   *Spin-Out Transition for Clip on Track A*

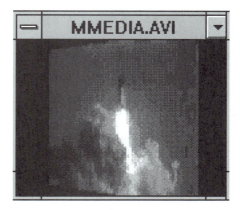

FIGURE 46   *Spin-In Transition for Clip on Track B*

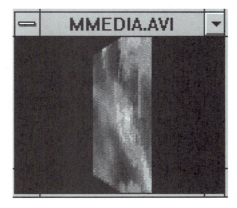

FIGURE 47   *Spin-In Transition Almost Completed*

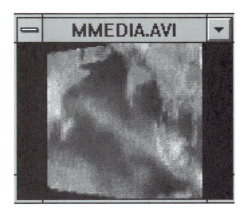

FIGURE 48   *Clip on Track B Plays to Completion*

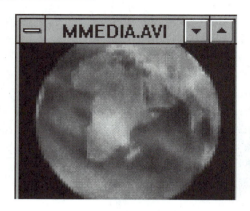

are Adobe's Premiere for Windows (for Video for Windows files) and D/Vision (for RTV files).

6. If you create your own movies, take steps to maximize the compressibility of your video. Many, if not all, camcorders include a power zoom feature. Do not use it for videos you intend to use on a computer. During the zooming process, most of the pixels in every frame are changing, which greatly diminishes the effectiveness of interframe compression. Have the camera firmly mounted on a tripod to eliminate jiggles (which also causes most pix-

FIGURE 49   *Creating the Movie File*

els to change from frame to frame), although software exists (e.g., from Doceo) that can eliminate most camera jitter effects.

7. Keep your clips short and to the point. Even compressed video takes up a lot of room, in the order of 9 megabytes per minute.

8. QuickTime movies created on the Macintosh can be played in Windows using Apple's QuickTime for Windows product.

9

Ancillary Devices

The devices described in this chapter play an auxiliary role in computer multimedia. If you get involved with multimedia, you will probably employ one or more of them.

Still-Video Cameras

If you would like to incorporate still images into your computer application, those images may be captured using a still-video camera. Still-video cameras, which came on the market in 1991, are capable of capturing black and white pictures as well as color. These cameras may be (a) digital, linking directly to your computer; or (b) analog, requiring the use of a video capture card in your computer (the same cards that are used to capture software motion video, described in Chapter 7). It is even possible to send images via phone lines. Some still-video cameras can connect a computer to a special still-video disc reader, which then allows the computer to send the picture through a modem.

The cameras are no more difficult to operate than a standard 35mm camera. Instead of film, a 2-inch diskette is loaded in the camera. You point and shoot to take a photo, just as with a regular camera. Pulling images from the still-video diskette will require a video floppy drive or a still-video player. However, the easiest way to bring the image into the computer is by connecting the camera itself to the computer. You will be able to select any of the pictures on the camera's diskette. Selecting, manipulating, and editing images can be done with software (see "Paint Programs and Image-Editing Programs" on page 95).

An example of a still-video camera is Cannon's ZapShot, which cost several hundred dollars and is now discontinued. This camera provides an NTSC output signal that can be connected to the video capture card in the computer. Several other digital cameras are available, although their cost is very high, in the order of several thousand dollars. An alternative, new to the market, is Apple's

QuickTake 100 camera, available for the Macintosh and for Windows. This camera holds 32 pictures of 320 × 240 resolution, or 8 pictures at 640 × 480 resolution. Pictures are in 24-bit color format. To bring the pictures into the computer, the camera is connected to the computer's serial port, thus eliminating the need for a video capture card in the computer. This camera features automatic focus and exposure, and thus is intended for consumer, not professional, use. However, depending on your photographic needs, it may meet your requirements.

As mentioned in Chapter 5, you might also consider using a high-quality 35mm camera, and have your photos written on a Photo CD.

Camcorders

If you are using a video-digitizing card, capturing your own video for integration into multimedia applications is possible with a home-type camcorder (Figure 50). In addition to the digitizing card, you will need special software to manipulate the video.

When selecting a camcorder, remember that they record at several levels of resolution. The better the recording level, the better your computer-captured video will be. From lower to higher quality, common camcorder formats are: VHS-C, VHS, S-VHS, Hi8, and digital Hi8. For example, VHS-C and VHS camcorders record at 240 horizontal lines per frame, while S-VHS and Hi8 record with 400 horizontal lines per frame. The highest-quality (and most expensive) cameras are the so-called RGB cameras.

Another consideration is the number of image-sensor chips the camera has. Most inexpensive camcorders have a single chip, which records the red, green, and blue components of light. More-expensive cameras have three chips, each one dedicated and optimized for an individual primary color. Needless to say, three-chip cameras produce better picture quality.

RGB cameras provide separate red, green, and blue signals (i.e., the signals that are generated by the red, green, and blue sensor chips) with very little processing. The S-VHS signal has separate luminance (brightness) and chrominance

FIGURE 50 *VHS Camcorder*

(color) signals. The consumer-level camcorders provide (in the United States) an NTSC signal, which mixes color and luminance information on a single wire. The signal has to be processed again in the capture card to separate its red, green, and blue components. Because the NTSC signal needs the most processing, it produces the lowest quality.

Finally, consider the importance of audio quality in your video. VHS sound quality is mediocre. VHS Hi Fi audio is much better. The 8mm camcorders have very good sound quality.

Camcorders share an important attribute with still-video cameras: They do not have to be connected to the computer at the time the picture or video is shot. So you can take your camera into the field, capture your material, and later connect the camera to the computer to transfer and digitize the material.

Scanners

Scanners provide another method for bringing images into the personal computer. They are used extensively to scan printed images, although they can also be used to scan photographs. Scanning is a powerful, easy, and fast way to incorporate images into any presentation or application.

FLATBED SCANNERS

Flatbed color scanners (Figure 51) allow you to scan whole images as well as pages of text. Before choosing a scanner, you need to consider several points:

1. Resolution. Resolution usually is defined in **dots per inch (d.p.i.).** This indicates the number of dots in one square inch of the scanned image. The minimum resolution should be equivalent to, or better than, the resolution of the device that will produce the image (e.g., the printer). Scanners often provide software routines that interpolate the pixels (i.e., create a new pixel between every pair of pixels by averaging the value of the adjoining pixels) in the scanned image, thus increasing the stated resolution. Optical resolution is the resolution provided by the scanner's image-sensing chips (called *charge-coupled devices,* or CCDs).

2. Interfacing. How does the scanner connect to your computer? Many scanners have an SCSI (small computer systems interface), so your computer must have a SCSI port. Others have a proprietary interface, so they come with a board that must be plugged into the computer. The software protocol is also important, so that the computer may control the scanner. Unfortunately, there are several different protocols for scanners on the PC. To resolve this problem, a scanner protocol standard, called **TWAIN (technology without an interesting name)** was developed and is becoming popular. Adoption of TWAIN means that more graphics programs can support scanning, as they only have to support one protocol.

3. Software requirements. Many graphics programs implement the TWAIN protocol, so they can interface directly with the scanner. If your graphics program cannot "talk" to the scanner, then you must make sure that the scanning program (which comes with the scanner) can save files in a file format that your graphics program can import.

4. Color capabilities. There are gray-scale scanners and color scanners. Most color scanners are 24-bit scanners, although scanners with higher color resolution, such as 30-bit scanners, are also available.

5. Speed. Scanner speeds vary greatly. Some scanners scan an image three

FIGURE 51 | *Flatbed Scanner*

times, one pass for each color (red, green, and blue), while others will scan all colors in one pass over the image.

6. Price. Speed is one factor affecting the price of color scanners. Generally, the speed is directly proportional to the price. Scanners with 24-bit color are available for less than $1,000. Image size also affects the price. Most color scanners accept images of up to 8 1/2 inches by 14 inches. Once you exceed these dimensions, the price dramatically increases.

HANDHELD SCANNERS

Another type of scanner is the handheld model (Figure 52), which can be useful for field applications (connected to a laptop computer, for example) such as scanning an image from a manuscript that cannot be removed from a museum.

Some handheld scanners can connect to your computer's serial port, others come with a special hardware card that plugs into the computer. Handheld scan-

FIGURE 52 | *Handheld Scanner*

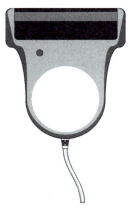

ners also come with a driver program, as well as scanning programs that can, for example, "knit" together a page scanned in several passes.

OPTICAL CHARACTER RECOGNITION

Scanners work by (what else?) scanning the page line by line and recording the color of each dot, then sending this stream of information to the computer. The scanned image is stored as a bitmap in one of several popular file formats, such as TIFF. As the bitmap has no "knowledge" (representation) of objects contained in the image, text that is scanned cannot be processed by the computer as text. To overcome this, a number of programs, known as **optical character recognition (OCR)** programs, exist that can examine a scanned image and recognize the text information contained therein. Some are even able to preserve character formatting so that the text may be pasted intact into a word processing or desktop publishing package. An optical character recognition program must have a very high recognition rate, or the effort that it takes to correct all of the recognition errors will offset the time savings gained by scanning, instead of typing in, the material.

Graphics Tablets

The mouse has proven to be a very popular input device on computers that feature a **graphical user interface (GUI),** such as Windows. A mouse lets you quickly move the pointer to any point on the screen, and the two (in the case of Windows) mouse buttons allow you to select objects, pull down menus, and so on.

However, the mouse is an inadequate input device when it comes to some graphics programs, as the mouse cannot reproduce the stroke of the artist. The impression you make on paper when using a pencil, for example, depends not only on the nature of the pencil and on the surface quality of the paper but also on how you move the pencil across the paper. You might use a light touch, leaving a faint line on the paper. Or you might use more pressure, resulting in a bold line.

Graphics tablets (Figure 53) are input devices that overcome this limitation. They consist of a flat surface (the tablet) and an electronic pen. As you move the pen across the surface of the tablet, the mouse pointer on the screen moves with it. The graphics tablet furnishes the computer with the position of the pen on the tablet, which the computer translates into the mouse pointer position on the screen. The tablet also provides an indication of the amount of pressure with which the pen is pressed against the tablet. Certain graphics programs, such as CorelDRAW! and Fractal Design Painter, are able to read this pressure information and transform it into realistic drawing strokes on the screen. Another advantage of a graphics tablet over a mouse is that it feels more natural for people to draw and paint with the tablet than with the mouse. You may also place a drawing on top of the tablet and trace it with the pen. Note that the pen on a graphics tablet does not draw anything on the tablet itself. The position and pressure of the pen are sensed by the tablet and transmitted to the computer.

The tablet connects through the computer's serial port. A graphics tablet's specifications will include drawing area, type of pen, resolution, and pressure levels. Some pens are connected to the tablet by a thin cable, others are wireless. The cost of the tablet increases with the size of the drawing area of the tablet. A small, 4×5 inch tablet, with a wireless pen and capability of detecting 256 pressure levels, with resolution of over 2,000 points per inch, can be purchased for much less than $200. Graphics tablets manufacturers include Wacom and Calcomp.

FIGURE 53   *Graphics Input Tablet and Pen*

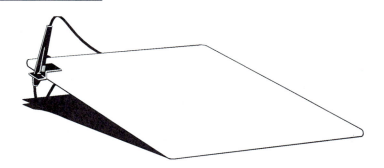

Multimedia Presentation Devices

For your multimedia presentation to achieve its maximum impact, you must take into consideration the technology used for classroom presentations. Classroom multimedia presentation systems generally fall into three categories: three-beam projectors, color LCD panels, and large-screen monitors.

THREE-BEAM PROJECTORS

Standard lecture-hall projectors are often referred to as "three-beam," "three-gun," or "RGB" (red, green, blue) projectors. They have three cathode ray tubes, one each for red, green, and blue light. It is preferable for these projectors to be permanently mounted in one place since convergence adjustments must be performed each time the projector is moved. These projectors are thus best suited for auditoriums or large lecture halls.

Features to note about three-beam projectors include:

- Overall brightness, or lumens output.
- Level of horizontal resolution—higher levels indicate a clearer picture.
- Horizontal scan frequency. To determine if a projector is compatible with your computer, compare the horizontal scan frequency ranges of both the computer and the projector. For example, VGA displays require projectors that operate at scan rates of 50 KHz.
- The maximum distance between the screen and the person farthest away is eight times the height (top to bottom) of the screen.

These projectors tend to be expensive, but provide the best picture quality. Many schools have these projectors permanently installed in special "multimedia" lecture halls.

In addition to video signals from the computer, some RGB projectors may also accept other signals (NTSC, standard television, PAL or SECAM the European equivalent, as well as S-VHS, which is generated by camcorders, VCRs, etc.).

COLOR LCD PANELS

Liquid crystal display, or **LCD,** panels (Figure 54) give you the ability to display information from a computer screen using a conventional overhead projector. The LCD panel can be connected to the computer and placed on top of an overhead projector as if it were a transparency. The transparent LCD plates inside the panel create an image that is displayed on a wall or screen with the optics of the overhead projector. These panels are more portable and less expensive than three-beam projectors, making them a perfect option for use in small- to moderate-sized classrooms. Some LCD panels include auxiliary inputs so that other video devices, such as a VCR, may be directly connected to them. This avoids the need to switch cables around during a presentation. Most panels also have remote controls, so you can turn off the display, switch input sources, or adjust the color without needing to touch the device.

Two types of LCD panels are currently available: **passive matrix** and **active matrix.** If you plan to project full-motion video, you will need an active-matrix (also called *thin film transistor,* or TFT) panel.

Passive-matrix panels turn pixels on and off one row at a time, which results in a refresh rate (how often the image on the panel is updated) that is too slow to effectively display full-motion video. So, while passive-matrix panels are more affordable, they are inappropriate for motion video.

Please note that this technology is evolving rapidly; some experts speculate that LCD panels will soon be capable of projecting images that are of equal quality to those of three-beam projectors. Below are some additional tips that may be helpful when looking for the panel that best suits your needs.

- All LCD panels require that the overhead projector produce a fairly high level of light. Most vendors recommend that the overhead be capable of producing 2,700 to 3,000 lumens. Many of the older overhead projectors available on your campus may not be powerful enough to use with LCD panels.
- One feature available on many panels is VGA-out, which allows the panel to send the VGA signal it is displaying to a separate VGA monitor. This feature is useful for presentations because it allows you to face the audience and view the projection from an auxiliary VGA monitor, while the audience views the projection at the front of the room.
- Displaying images on a wall is not uncommon, but it is strongly recommended that you purchase a high-quality screen. Most screen manufacturers are now marketing a screen called "Spectra Surface" that is specifically designed for use with LCD panels.
- You need to be able to control the lighting level in the room, or it will be difficult for the audience to see the projected image. This includes not only the ability to dim or turn off the lights in the room but also the ability to control light from external sources, such as daylight coming through the room's windows.

FIGURE 54   *Color LCD Panel*

LCD Projectors   A newer technology combines the features of the LCD panel with those of the three-beam projectors. These devices have their own built-in light, which shines through an internal color LCD panel. They have only a single beam, so the only adjustment necessary is focusing. LCD projectors provide a much brighter image than the overhead projector and LCD flat panel combination, and are more convenient because there is only one device to carry. However, they are also more expensive than LCD flat panels.

Large-Screen Monitors   Multimedia applications may also be presented using large-screen monitors. The screens can measure from 21 to 36 inches diagonally. Although these monitors provide excellent image quality and are relatively portable (both computer and monitor, plus peripheral devices, can be placed on a cart and wheeled from one classroom to the next), they may be too small for large audiences.

Several features should be considered before purchasing a large-screen monitor:

- Be sure to determine if a particular model accepts VGA signals. Standard televisions are inappropriate for anything but regular video.
- Many large-screen monitors hook directly to the computer; some require a converter.
- The lower the "dot pitch" number, the better.
- Nonglare screens are obviously better.
- The higher the screen resolution, the better the image quality. Screen resolution can range from 640×480 up to $1{,}280 \times 1{,}024$.
- Multisync monitors are available that automatically match the output of the video card installed on your computer so that you need not change settings if you switch to a different computer.
- Allow 2 inches (diagonal measure) of screen for every person in the audience. Thus a 30-inch screen monitor can accommodate an audience of 15 people.

Regardless of the projection system used, test the legibility of the text on the screen.

10

Networking

We have seen that multimedia files are very large, even after compression is applied. This provides a strong incentive to place these files on a network **server,** so that massive amounts of identical data are not duplicated on each computer in a lab or on multiple computers on a single campus. It is also desirable to network the material on videodiscs, as obtaining an individual videodisc player for each computer is expensive and takes up a lot of room. In this chapter we will discuss how multimedia, in both digital and analog format, may be networked.

Analog Networking

We have seen the use of CD-ROM servers on a network (page 42). Wouldn't it be nice to do the same with laser videodiscs? Then perhaps ten players could service a lab of 30 computers, because not every student would be accessing video all the time. The problem is that the video signal provided by the player is analog, while our networks are digital. Some people solve this problem by using a computer to digitize the video signal as it comes off the disc and then placing the digitized video on a conventional digital network. This approach has problems of its own, as we will see when we examine digital networking.

A popular solution is to run the video on a separate, analog network connecting the servers to the student's computers. Only one coaxial cable is needed for the network, as multiple video signals can share the cable's bandwidth, exactly in the same way as is done on cable TV. At each student station a tuner, such as IBM's PS/2 TV, is electronically tuned to the correct channel; the tuner retrieves the signal off the network and places it on the student's screen, alongside the computer-generated screen graphics and text. In detail, this is how it works:

1. The program running in the student's computer reaches a point where a specific video sequence is called for. The program sends a short message through the regular digital network requesting the video.

2. The server receives the message and consults its tables to see which videodisc players have that specific videodisc mounted. It selects a player that is not currently busy.

3. The server sends a message to a **multiplexer** to place the output signal of that player on an unused channel on the cable. The server also sends, through the network, a message to set the user's tuner to that same channel.

4. The server sends a command, through a serial interface, to that particular videodisc player to play the sequence of frames desired. It also updates its tables to show that player as busy and that channel as in use.

5. The player plays the sequence, which appears on the user's screen.

6. When the sequence is completed, the server updates its tables to show the new status of player and channel.

If the disc is not available, the server can alert an operator as well as inform the student, through a network message, of this condition. If the disc is in a player, but the player is busy servicing another request, the user can wait, or the server can relay an error message.

This kind of implementation has a number of advantages. For example, with some networks, the video might be combined on the same cable as the digital network, saving the cost of another cable. PC VCRs might also be used, where students watch long sequences of videos. The only difference is that the student is watching from a computer screen instead of from a TV monitor. The same cable might also carry satellite broadcasts, campus TV channels, and so on. The amount of processing required from the server computer is minimal, so it can be used for other purposes as well, perhaps as a file or print server.

These analog networks are in use. One example is the VIDS network at the University of Michigan. Another example is the Ultimedia Video Delivery System/400 from IBM, which utilizes an IBM AS/400 system as the server.

Digital Networking

As digital video gains in quality and popularity, the demand to serve these files across a network will increase. A significant advantage of, for example, software motion video technologies is the absence of any specialized hardware requirements at the user's computer.

It would seem that meeting this demand for digital networking would not present a problem. Network technology is well developed, and network speeds are high: 10 megabits per second for **Ethernet,** 16 megabits per second for token-ring networks (networks that use a token-passing scheme to control access), and even higher speeds for **ATM (asynchronous transfer mode)** networks (see page 88). Surely we can transfer a Video for Windows file with a transfer-rate requirement of 200 KB/sec (1.6 megabits per second) over these networks without a problem. In reality, however, although utilization (average amount of data transferred) of most networks is low compared with their capacity, multimedia files present special problems.

To understand these problems, let's consider how a **LAN (local area network)** is used. One common use for these networks is for storing programs that are to be used by multiple students. There might be a networked lab on campus containing a word processing program. Any student on a computer connected to the network can load the word processor and run it. When the student loads the word processor, parts of the word processor program are read from the server's disk and copied into the memory of the computer the student is using. The whole program is not copied; for example, the parts of the program that perform spell checking, printing, or other services, are not loaded until and unless they are invoked.

The server reads portions of files and places them on the network. These pieces are not necessarily in order on the disk, and some pieces will only be requested much later, if at all. The design of the network is optimized for this kind of use.

In contrast, let us now consider the use of a 30-second Video for Windows video. The file is 6 megabytes in size, or larger. It will probably be transferred in its entirety, and in order (sequentially). Furthermore, the video should start playing as soon as the first frame arrives at the student's computer. To wait for the whole video to be transferred before it starts to play would impose unacceptable memory requirements on the student's machine, as well as result in long delays from the time the movie was invoked until it started playing. Once the movie starts playing, then, the network must deliver the frames at the required rate and without interruption. This kind of delivery is called **isochronous.**

A student will not notice if the word processing program takes 1/4 second longer to load today than it did yesterday. However, the same student will definitely notice a delay of 1/4 second in the middle of a video stream, as it will cause several frames to be skipped. Yet such delays are common, even in lightly loaded networks, because the network might be delivering data to another computer. Even if we make our networks faster, there is always the possibility that at a crucial moment the network will not be able to deliver a sequence of frames in time.

In practice, it has been impossible to support more than two or three multimedia users on a network; this is not an affordable solution, because theoretically, a token-ring network, for example, should have the capacity to transfer up to 10 videos at once.

Several technologies have emerged to address this problem. Products are available from a number of vendors, such as Fluent (purchased recently by Novell), Protocom, Starlight, and IBM (IBM LAN Server Ultimedia). To illustrate how the delivery problems are solved, let's look more closely at one of these products.

TECHNIQUES FOR DELIVERING MULTIMEDIA OVER A DIGITAL LAN

The IBM LAN Server Ultimedia product can be installed on an existing or new IBM LAN Server network, which services DOS, Windows, OS/2, and Macintosh machines. The same network handles both conventional and multimedia traffic. The product has the capacity to transfer up to 40 different video files simultaneously from a single server, over four token-ring network segments. It also has a quality-of-service guarantee. If the server accepts the request to transfer a multimedia file, the data is guaranteed to arrive at the user's workstation in a timely manner with no interruptions. This guarantee applies only to token-ring networks.

How is this done? Several techniques are employed in parallel to be able to achieve such transfer requirements.

1. The way the server accesses and buffers files is changed. Once the server receives the request for the file, it reads as much of the file as it can into its memory (the buffer), so that it will have this data ready to place on the network when needed.
2. The network administrator can set limits to both multimedia and conventional data transfer rates. When the server starts to transfer a multimedia file, it reserves the required bandwidth on the network so that at no time during the transfer will the network be too busy to deliver the data.

3. Multimedia data is sent with a priority token, so it takes precedence over conventional data traffic on the network. Prioritization is not possible on Ethernet networks, which is why the quality-of-service guarantee applies only to token-ring networks.

4. The server is aware of multimedia file characteristics. It knows the transfer requirements for each type of multimedia file, and uses this information to manage its resources.

5. The server knows, through testing, the data transfer rate of the disks (and their controllers) that have multimedia files. Thus it can schedule disk activity so that, once a multimedia transfer commences, at no time will the disk be too busy to continue providing the file's data at the required rate.

How does all this multimedia traffic affect the delivery of regular files over the network? Not as much as you might expect, because on average, traffic on local area networks is light. Benchmarks were run on a local area network running LAN Server Ultimedia. A 2 KB file, which took 0.11 seconds to transfer when there was no multimedia traffic on the network, took 0.12 seconds to transfer when the network was transferring ten multimedia files. A 1-megabyte file, which took 1.73 seconds to transfer when there was no multimedia traffic on the network, took 2.36 seconds to transfer when the network was transferring ten multimedia files. A measurable increase, but not one that would unduly inconvenience the user.

The Future of Multimedia Networking

Even though the above solutions cannot guarantee quality of service on an Ethernet, other approaches are available. One is to use intelligent Ethernet hubs, which are star-wired to individual computers (i.e., each computer is directly connected to the hub). In essence, this gives each computer its own "private" Ethernet, so multimedia data is not disrupted. The intelligent hub manages the overall traffic, so the load on each Ethernet wire never precludes the transfer of a multimedia file. Typically each Ethernet station is guaranteed an isochronous (see page 86) transfer rate of 6.144 megabits per second. The intelligent hub is connected to a server through some other high-speed network. The drawback of this approach is the cost of the intelligent hubs, and the need to rewire existing Ethernets.

Current token-ring networks run at 16 megabits per second. Other network technologies exist, such as **FDDI (fiber distributed data interface),** with a transfer rate of 100 megabits per second. Using the techniques discussed on page 86 on these other network technologies increases the number of simultaneous multimedia file transfers on a single network. FDDI is used as the backbone network on many university campuses. FDDI uses fiber optics as the transmission media, but FDDI derivatives, such as **CDDI (copper distributed data interface),** are becoming available that use standard copper wire and achieve the same transfer rates.

It is all well and good that you are able to set up a lab with a single server and with multimedia computers and serve them without problems. However, the problem also applies on a much larger scale. Suppose you are preparing a lecture and need some material that a colleague has available across the country. If the information is in printed form, it can be mailed or faxed to you. What if it is image data? Perhaps your colleague can download the data from the local server and copy it to diskettes for you, or send it across the Internet. But what if the information is the recording of a speech, or a video of some event taking place? Perhaps a video or an audio tape may do.

Let's speculate further. You have prepared and presented a sophisticated multimedia presentation that has inspired your students and generated a lot of classroom discussion. A comment or a question from a student causes you to recall that there is a video clip that would be most appropriate to show right now, while the discussion is in progress. Unfortunately, that video clip resides on a computer in another city or state. Sure, you can download it or have it sent to you later, to show at another time, but by then it might be too late to have as much impact. What you would like to do is to call up the video now, and show it.

This may very well be possible within a reasonable time frame. ATM (asynchronous transfer mode), a new network technology, is just arriving on the marketplace. ATM is well suited for isochronous data transfer, and is implementable across wide area networks. It supports transfer rates from 45 to 622 megabits per second. Two characteristics of ATM are small block (the amount of data transferred as a unit) size (40 bits of header information, containing address, control and error-checking information, and 384 bits, or 48 bytes, of data) and very fast switching of the ATM circuits (150 microseconds), which means it takes the network very little time to switch from transferring a block of your data to transferring a block of someone else's data. The small block size, 53 bytes, guarantees that the circuit cannot be tied up long with the transfer of any single block. The hardware technology and the protocols defined for ATM make it possible to guarantee, at a high probability level, isochronous delivery of data. Commercial carriers are already implementing ATM networks connecting major metropolitan areas. Local area networks based on ATM are also currently available. IBM, for example, has 25-megabit ATM cards on the market.

Think of the possibilities brought about by having enormous amounts of scholarly multimedia data available at your fingertips almost instantly. If you are planning for a project that will be implemented in the next few years, you will be able to take advantage of the ability to transmit multimedia across local and wide area networks.

The Internet

The **Internet** is quickly becoming the network of choice for communication by computer for people all over the world. As the name implies, the Internet is not a single network but a network formed by connecting thousands of networks together. There is a vast amount of multimedia material available on the Internet. Some examples include:

- Pictures of artwork from the Smithsonian Institution (Washington, D.C.) and the Louvre Museum (Paris), to name but two.
- Photographs taken by the Hubble Space Telescope.
- Current weather maps (Figure 55), radar plots (updated hourly), and satellite photos of cloud cover (both visible and infrared) over the United States and the world.
- Video clips of news and historic events.

HISTORY

The Internet was not planned but rather it evolved with the needs of its users. The Internet evolved from ARPAnet of the '70s, which was an experimental network intended to support military research. In order to assure reliability, the network was designed with flexible routing, so transmissions over the network would arrive at their destination even if part of the network was not running.

FIGURE 55   *Weather Map Available over the Internet*

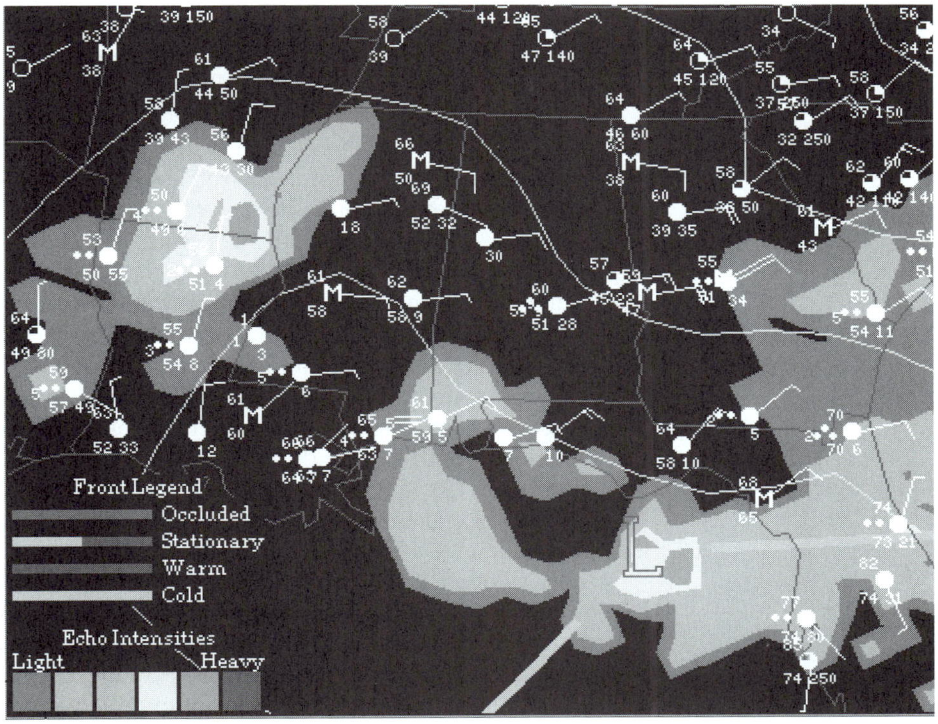

In the '80s, UNIX machines began to be connected on networks, and some of these networks were connected to each other over the ARPAnet. Then the National Science Foundation funded the establishment of five supercomputing centers around the country, and the supercomputers became available to researchers remotely over NSFnet, which was a descendant of ARPAnet. Researchers could submit their programs over the network to be run on the supercomputers.

Soon the supercomputers' capacity became saturated, and researchers turned to other computers connected over the network to get their work done. This has evolved to the point that in September of 1994 there were 2.8 million host systems on the Internet.

Here are some 1994 statistics that give an idea of the growth of the Internet:

- A new network is connected to the Internet every 10 minutes, on average.
- **E-mail** reaches 137 countries over the Internet.
- The annual rate of growth for **Gopher** traffic on the Internet is 997%.
- The annual rate of growth for **Mosaic** traffic on the Internet is 341,634%.
- There were more than 3,000 newspaper and magazine articles about the Internet in 1993.
- In November 1994, there were 39 elementary schools that had **home pages** (see "Mosaic" on page 91) on the Internet.

GOPHER, ARCHIE, AND VERONICA

Early Internet tools, such as **TelNet** (remote login) and **FTP (file transfer protocol),** were developed for UNIX systems. The command language is not simple for those used to graphical user interfaces such as Windows. In addition, although there was a wealth of resources on the Internet, it was not easy to find anything if you did not know where to look.

For these reasons new tools were developed, including Gopher, **Archie,** and **Veronica.** Gopher is a menu-based navigation system that lets you easily navigate the contents of a Gopher site. The site (a computer) puts up a Gopher server program, and you run a Gopher client program on your computer. Gopher was developed at the University of Minnesota; its name not only reflects the name of Minnesota's mascot (the Golden Gophers) but is also a form of "go for," or "go fer"; in other words, the program retrieves the information the user wants.

To obtain Gopher you must pay a fee to the University of Minnesota. However, there are several Gopher-like software packages available for free over the Internet.

There are many Gopher servers on the Internet (well over 1,000). Thus finding the one that has the information you want can be time consuming. For this reason, tools such as Archie and Veronica were developed. Archie, named for the president of the university where it was created (McGill University), is an "agent" that goes around the network searching for a file you want and sends it back to your computer when it finds it.

Veronica (for *very easy rodent oriented netwide index of computerized archives,* although one suspects the name was inspired by the comic book characters Archie and Veronica) is a tool that lists what is available at Gopher sites. It was developed at the University of Nevada, Reno. Figure 56 shows a list of Veronica sites, found using Gopher client software. Notice the "menu" organization of the results—this is a typical Gopher screen. Also notice that sites are located in different countries. When running Gopher, clicking on a menu item might take you to a system halfway around the world without you even noticing it.

Suppose we are searching for the text of a sonnet by poet Edna St. Vincent Millay. After selecting a site (any site will do), type in the search words (Figure 57).

Veronica returns two menu items (Figure 58) as a result of the search. They are probably located in a system other than the one that performed the search. It does not matter. Simply by clicking on one of the items, Gopher will retrieve the file and display it on your screen.

FIGURE 56   *Veronica Sites*

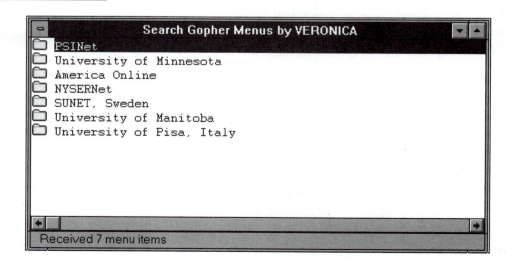

FIGURE 57 | *Specifying the Search*

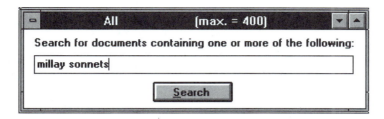

WORLD WIDE WEB

The people at CERN (European Laboratory for Particle Physics) in Switzerland developed a way to tie information together among different systems all over the world. This technology is known as the **World Wide Web (WWW).** Documents are created that have hypertext or hypermedia links to other documents, or multimedia files, that can be on any other system on the Internet.

MOSAIC

The **National Center for Supercomputing Applications (NCSA)** at the University of Illinois developed a graphical program, called *Mosaic,* which allows you to navigate the World Wide Web documents.

A web starts with a home page, which is a document that has hyperlinks to other documents or multimedia files. Figure 59 shows a portion of a Mosaic screen. Notice the use of text formatting (different fonts). The page could also include graphics, movies, and so on. The words in blue are *triggers.* Clicking on them will take you to a linked document, which might or might not be on the same system.

There are several World Wide Web browsers in addition to Mosaic, including Cello, developed at Cornell University, Netscape, and Web Explorer.

FIGURE 58 | *Results from Veronica*

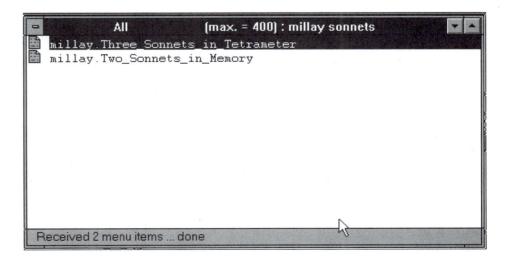

FIGURE 59   *Part of Mosaic Window*

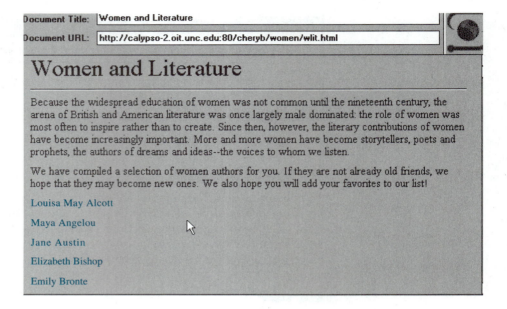

The latest development over the Internet is the availability of virtual reality clients and servers. These let you navigate by visually moving through connected rooms, very much in the way one navigates through popular computer games such as Doom.

11

Software

A discussion of multimedia technology and hardware would not be complete without a survey of software programs that enable you to put multimedia to use. In this chapter we will take a close look at a number of program categories and representative products. This will be useful if you are starting out and need to know what is available.

The Role of Standards

A few years ago, the author of a multimedia program would have to select which multimedia devices to use and then write the program to directly control those devices. Another author might want to use the same devices, which required that his or her program also incorporate the necessary function to control the device. Suppose both programs employed a certain audio card. If now a third author were to write a program, but chose a different card, the program sections that interfaced with the card would have to be different from those in the first two programs, because the card would have a different command set (the set of commands it understands) and protocol from the first audio card. Furthermore, this third program would not be able to share the sound files with the first two, because each card defined a different digital representation of sound.

The adoption of standards simplifies the task of the author and of the end user. If each audio card, for example, responds to a common command set and syntax, then the author of a program does not need to worry about which audio card the user might have, or have to supply multiple drivers with the program to support a variety of cards. If the data formats are standards-based, then the author knows the sound files supplied with the program will work on a variety of machine configurations. The end user can purchase the software with the security of knowing that it will run on his or her computer. The enormous popularity of the IBM PC and compatibles is partly due to the fact that they all adhere to an informal standard, so that a single software program will run without modification on any of them.

MULTIMEDIA STANDARDS

We have already talked about several standards as they apply to multimedia, such as those that govern the format of CD-ROM discs and devices (Chapter 5). Many of the multimedia file formats are also standardized, such as WAV files (page 47) and general MIDI files (page 51).

MPC MPC (Multimedia Personal Computer) is a standard that specifies the hardware requirements and functional requirements (such as CD-ROM access time, waveform audio sampling rates and sample sizes, etc.) for a computer to be multimedia enabled. This standard is controlled by the MPC Marketing Council, Washington, D.C. Many personal computers, hardware accessories, and software products carry the MPC seal, which provides a measure of assurance that the pieces will work together. Table 5 and Table 6 detail the hardware requirements for MPC-2 compliance, which is the latest level of the specification.

MCI MCI (Media Control Interface) is a specification implemented in Microsoft Windows that provides a standard syntax and protocol for controlling and sharing all multimedia devices. Programs control these devices by issuing simple commands to the MCI interpreter, a program which is a part of Windows. For example, to start playing a CD-ROM audio disc, the program needs to issue only two commands, as follows:

OPEN CDAUDIO

PLAY CDAUDIO

The syntax of the MCI language is simple and consistent. For example, all multimedia devices, be they an audio card, a MIDI synthesizer, a CD-ROM drive, a videodisc player, and so on, support the concept of playing. Thus one can issue the PLAY command for any of these devices by simply sending the string:

PLAY device_name

(where device_name gets replaced with the actual name of the device) to the MCI interpreter. While simple, the language provides for comprehensive control of the devices. For example, one can specify the precise starting and stopping

TABLE 5 | *MPC-2 Minimum Requirements*

Item	Minimum Requirement
RAM	4 megabytes
Processor	25 MHz 486SX
CD-ROM drive	300 KB/sec sustained transfer rate, maximum average seek time 400 milliseconds, CD-ROM XA ready, multisession capable
Sound	16-bit digital sound, 8-note synthesizer, MIDI playback
Video display	640 × 480, 65,536 colors
Ports	MIDI I/O, joystick

TABLE 6 | *MPC-2 Recommendations*

Item	Recommended Support
RAM	8 megabytes
CD-ROM drive	64KB on-board buffer
Sound	CD-ROM XA audio ability, support for IMA-adopted ADPCM algorithm
Video	Delivery of 1.2 mega pixels/sec given 40% of CPU bandwidth

points for a video sequence from a videodisc player by issuing a command similar to the following:

PLAY VIDEODISC FROM aaaa TO bbbb

where aaaa and bbbb represent the starting and ending frame numbers.

Media-less devices (i.e., devices that get their data from files on the computer's disk or from across a network, rather than from physical media such as CDs or laserdiscs), such as an audio card, require that the program specify the name of the file to be played with the OPEN command. Commands are also defined to allow a program to ascertain the status of a device (playing, stopped, current frame number, etc.), to cause a device to notify the program when the multimedia action has completed or has reached a certain point (say, a specific frame number), and to ascertain the capabilities of the device (can it play 44.1 KHz WAV files, for example?).

There are two significant advantages to MCI. The first is that a software vendor can write a program incorporating multimedia by using the MCI command set. The vendor does not have to be concerned about the details of the multimedia device, nor about which brand or model of multimedia device the purchaser of the program might have. Likewise, the device manufacturer can design a new device (such as an improved audio card) and know that users will be able to install the device and use it with their software. This is illustrated in Figure 60.

The second advantage of MCI is that control of multimedia devices is made easy, which also encourages the incorporation of multimedia into programs. It is also very easy to incorporate support for additional device types. For example, the commands required to play an animation file, a digital video file, an audio file, or a MIDI file are exactly the same. The only change is the name of the file itself.

All of the newer programs for Windows with multimedia capabilities make use of the MCI interface. Many authoring programs also make the MCI interface accessible to their users.

Multimedia Software

There are so many software categories that it would take a book to describe each one. For the sake of brevity we will only cover the major categories, and only list a few representative products in each category. The exclusion of a product does not mean that it is less capable.

PAINT PROGRAMS AND IMAGE-EDITING PROGRAMS

Programs in this category are used for creating and editing bitmaps. They allow editing of the image pixel by pixel, if necessary. Once an element is placed on the image, it cannot be manipulated as an entity. The simpler programs in this cate-

FIGURE 60   *MCI Illustration*

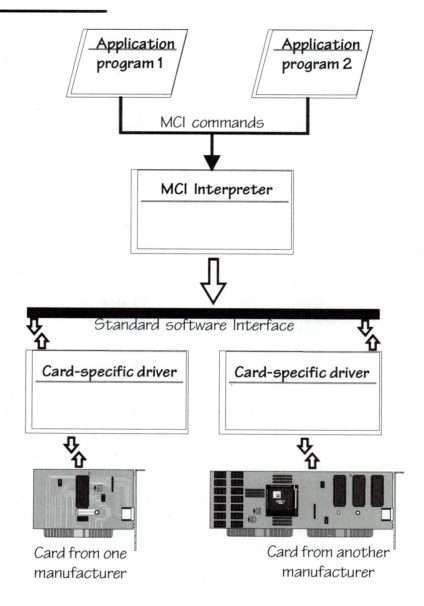

gory provide tools for drawing shapes and lines of many colors, and for filling shapes with a variety of colors and patterns. Advanced programs try to duplicate the function of a real canvas, having tools corresponding to the various artistic drawing and painting materials available, such as pencils, charcoals, chalk, water-colors, oil paints, crayons, and so on. They also realistically simulate the effect of the drawing material on a variety of user-selectable canvas textures. Some support input devices, such as pressure-sensitive tablets, that lend themselves better to the drawing process than does the mouse. Programs that provide for image editing allow you to change an image's sharpness, tonal balance, contrast, and brightness. These effects might be applied to the whole image or to a selected portion of the image. Special-effect filters (e.g., posterizing) are included. Some programs are able to control scanners to bring in the material in the first place. Others even supply brush tools that imitate the strokes of famous artists such as Van Gogh.

In reality, only people with artistic capabilities will be able to make full use of these programs. However, they are still useful to the casual user for scanning in material and for touching up and improving the appearance of images. Most casual users will import most of their image needs instead of creating them from scratch. Even images scanned with 24-bit scanners will probably need some color correction, which these programs can provide.

Windows Paintbrush This is a very simple program that is included with Microsoft Windows. It supports BMP and PCX file formats. It is capable of simple painting but not image editing. The advantage is that it is free.

CorelPHOTO-PAINT This program is part of the CorelDRAW! product. It supports 24-bit color, having multiple images open at the same time, scanner support, tonal and other image transformations, special-effect filters, and color separations (printing an image by printing each of the primary colors on a separate page). This is a very capable program equally at ease as a paint program and as an image-acquisition and editing program. Many file formats are supported, including JPEG and Kodak Photo CD. The latest version of CorelPHOTO-PAINT also supports Adobe Photoshop plug-ins (below).

Adobe Photoshop Photoshop was probably the first sophisticated image processing program to become available, initially on the Macintosh, but now on Windows as well. Many university art departments will not consider any other product. However, some of the innovative features that first appeared in this product are now available in some of the other products described here. Photoshop provides precision and flexibility. In addition, it provides a standard interface for third-party vendors to plug in their own software tools as if they were a part of Photoshop itself. These plug-ins might be special-effect filters, a driver to support a scanner, and so on.

Fractal Design Painter This is probably the most unusual program of the group (you will suspect this the first time you see it; unlike other software products, which come in boxes, Fractal Design Painter comes in a paint can), and the one that most closely mimics the artist's tools and canvas. It has a large variety of artist's tools (all of the ones mentioned above, plus airbrushes, colored pencils, ink pens, felt pens, etc.) Each tool has a large number of possible adjustments, and you can create your own tools. The program supports 24-bit color, bringing in images from a scanner, and a variety of file formats. Artistic effects include Van Gogh, Seurat, Impressionist, and Flemish rub. Images, or portions of an image, can be rotated, distorted, filtered, and so on. The features are just too numerous to mention. Text entry is supported by the use of friskets, which are masks that can be applied over the painting to protect certain areas from being painted over. Thus you create a text frisket, position it, and then use one of the other drawing or painting tools to paint the text through the holes in the mask.

Fractal Design Painter is the first program to break the rule that paint programs cannot deal with painted elements as objects. In September 1993, Fractal shipped the X2 extension to the product, which allows you to paint floating elements. These elements look no different from others, but they can be selected as a unit, erased, moved, and even placed in front of or behind other objects. Thus the artist can paint a tree in a scene, and if it does not look right, the artist can move it somewhere else on the canvas or erase it. This is a standard feature of drawing programs, but it is new for paint programs. Other programs, such as CorelPHOTO-PAINT and Adobe Photoshop, are now also including some form of object manipulation. The latest version of Painter incorporates the X2 extension into the main product.

Choosing a Product   If you are experienced in art, you probably already have a favorite program. If you are just starting out, begin by using the Paintbrush program that comes with Windows. It might meet your needs, and it is simple enough to be quickly mastered. You may decide that you would rather not fool around with painting programs; or if you do want a more sophisticated product, you will have a better idea of which features to look for. If your needs include scanning and retouching the scanned images, CorelPHOTO-PAINT is an excellent choice if you have a need for the other components of the CorelDRAW! set. If you are interested in professional-quality photo retouching, Adobe Photoshop might be your best choice, although CorelPHOTO-PAINT is not far behind. If you need to create realistic-looking paintings of different media, using the computer and input tools such as a pressure-sensitive tablet, Fractal Design Painter is an excellent choice.

If you need an excellent painting program but Fractal Design Painter is too complex or too expensive for your budget, the Dabbler program (same vendor) might be your best choice.

DRAWING PROGRAMS

Unlike paint programs, these programs create drawings by placing drawing elements on the screen. These elements can be individually manipulated and customized. These programs come with large collections of useful clip art, and are very adept at handling text. Objects can be filled with colors, patterns, multi-color blends, and so on.

One attribute that sets these programs apart is their transformation capability. They can fit text to a curve (see Figure 7 on page 12) be it an arc or an irregular curve. Text can be made to fill an irregular-shaped "envelope." Objects can be extruded to give them a three-dimensional look and to give the illusion of perspective. A light source can be defined and the program will shade objects accordingly (however, not with the degree of realism achieved by specialized 3D rendering programs).

Representative programs of this genre include CorelDRAW!, Gold Disk Professional Draw, Arts & Letters Editor, and Micrografx Designer.

CorelDRAW!   CorelDRAW! is the leader of the field. It provides a wealth of features, yet it is not too complex to master. The current version also includes the CorelPHOTO-PAINT program, a business graphics program, an animation program, a presentation program, and a desktop publishing program, among others. Even at list price it is an excellent value. The product is shipped on two CD-ROM discs, and features several thousand pieces of clip art and over 800 typefaces in Adobe Type 1 format and TrueType format.

CorelDRAW! has been updated every year. Prior editions are still available for sale at reduced prices. This might be an excellent buy for a person who occasionally needs to add rotated text to a program, or to fit text to a curve. The clip art and font collections, as well as all the other programs that come with the drawing program, are bonuses.

Choosing a Product   First, ascertain your need for such a program. Many authoring and presentation programs include drawing tools of their own. However, they cannot achieve the range of effects made possible by the drawing programs, such as fitting text to any curve, distorting text to fit an envelope, perspective effects, and so on. Of all the programs, CorelDRAW! dominates the marketplace and has consistently received very positive reviews in computer publications.

Perhaps your needs are more modest, and you don't want all of the fancy features that CorelDRAW! and its competitors provide. Or perhaps you need to create diagrams, flowcharts, and so on. Several programs have emerged in the last two or three years to address these needs. These programs are less expensive and easier to use than the full-featured drawing programs. They also share some unique capabilities, such as preserving the connection of lines in a diagram, flowchart, or organizational chart, even as the elements of the chart or diagram are moved within the drawing. These programs also ship with collections of "smart" clip art, that is, clip art that is predesigned to be interconnected with other clip art, minimizing the amount of drawing the user has to do. These programs allow those who are not artists to create good-looking diagrams and charts. Programs in this category include Visio, Aldus Intellidraw, and CorelFLOW.

WAVEFORM-EDITING PROGRAMS

These programs allow you to visually edit WAV files, as well as to splice files together, to change the volume of any portion of the file, and to apply special effects such as fades and reverberation. Waveform-editing programs also provide recording functions.

Wave for Windows   This program is a product of Turtle Beach Systems, and it provides a comprehensive set of editing functions. Any part of a waveform may be selected, cut, and pasted as if it were a drawing. Up to three separate audio files may be mixed into one. Special effects include fade-in, fade-out, and playing a file backwards. You can equalize the sound (boosting or cutting the volume of specific frequency ranges within the signal, such as bass and treble controls do). You can stretch or compress the duration of the sound without affecting its pitch, which is quite useful in getting the duration of the audio to match the duration of another component, such as an animation. We met Wave for Windows in the section "Working with Digital Audio" on page 48.

Choosing a Product   Most people will use only prerecorded audio, or will record their audio and leave it as is. However, if you really do have a need to edit WAV files, Wave for Windows is a good choice, but not the only choice.

MIDI PROGRAMS

There is a good variety of these programs available. They allow you to construct, edit, and play MIDI files. Some programs use their own MIDI file formats but are able to convert back and forth between their format and standard MIDI file formats.

Master Tracks Pro   This program, from Passport Designs, will probably satisfy the needs of most users. It allows you to do just about anything with a MIDI file, from writing new music to extensively editing an existing composition. It includes sophisticated processing options, such as the ability to increase or decrease the tempo of a passage smoothly over a time span.

Trax   This is another program from Passport Designs. It is simpler, less expensive, and has fewer features than Master Tracks Pro.

Choosing a Product   Unless you are a professional or amateur musician, you will probably have no need for a MIDI editing program, even if you use MIDI songs in your application. If you do want such a program, the ones above will serve you well. Many other fine MIDI editing programs are available. For example, Turtle Beach Systems sells a set of MIDI programs, which includes MIDI Tuneup and MIDI Session. The former is for those who are not musicians. It allows you to set the instrument for each channel, control channel volume, pan, reverberation, and so on, and to solo or mute any channel or combination of channels. The latter program shows the song in musical notation and allows you to edit the composition itself (add or delete notes from the song).

ANIMATION PROGRAMS

Several capable animation programs are available. They vary in their degree of sophistication and ease of mastery.

Macromedia Director   The best-known animation program is Macromedia Director (formally Macromind Director). It has a comprehensive set of animation capabilities, including the ability to animate text and to animate several objects at the same time. The program includes a scripting language, called *Lingo,* that allows animators to create sophisticated animations. This program had been a Macintosh exclusive for many years; however, it has recently become available for the Windows platform.

This is not a program for the casual user. It takes time and effort to master and requires a robust hardware platform for authoring.

Animation Works Interactive   This is a product from Gold Disk. It runs under Windows, and creates animations that can be played by other Windows programs. Although not as sophisticated as Macromedia Director, it is easier to learn, and will satisfy most simple animation requirements.

Add Impact!   This is another Windows animation program by Gold Disk. It is less comprehensive than Animation Works Interactive, but it allows programs that support OLE (object linking and imbedding) to play its animations even if they do not provide their own animation capabilities.

Autodesk Animator Pro   This is a DOS program developed by Autodesk, the makers of the Autocad program. It is a popular program whose animation files play in Windows. A number of Windows presentation and multimedia programs ship with the driver needed to run these animations in Windows, and with sample animations created by the program.

CorelMOVE   This program is another one in the suite of programs that accompany CorelDRAW!. It includes path animation and multiple-cell animations. It also has "morphing" capability; this is a sophisticated form of animation in which one image evolves, frame by frame, into a different image. CorelMOVE animations may be saved by the program as Video for Windows movie files, which allows you to add a number of effects to the animation using movie-editing programs. We met CorelMOVE in the section "Animation Programs" on page 32.

Choosing a Program   First evaluate the animation capabilities of the presentation or authoring software you use. They might very well meet your needs. If you need a separate animation program, you will find that Macromedia Director is the most sophisticated. The Autodesk product is capable and well supported in Windows. The Gold Disk products are the easiest to use and might meet your needs. If you need CorelDRAW! for any reason, then your needs will probably be well met by CorelMOVE, which comes with CorelDRAW!.

PRESENTATION PROGRAMS

This is a broad category that includes programs designed to create 35mm slides, transparencies, or on-screen presentations (we will only consider the on-screen capability here). They are capable of creating bulleted lists and building a variety of business charts. Some include outliners, and a few have hypertext capabilities, so the presentation need not proceed linearly through the set of screens. There are many programs in this category that run under Windows, such as Microsoft PowerPoint, Asymetrix Compel, Aldus Persuasion, Micrografx Charisma, and Harvard Graphics. The field also includes some specialized products, such as Stanford Graphics, which excel in presentations including sophisticated scientific graphs. All products in this category provide a variety of screen transition effects and can print speaker notes and audience handouts. If you find a mistake in your presentation two minutes before it is to be delivered, you will still have plenty of time to correct it.

One of these programs might be appropriate for you if you are just getting started, want to do multimedia presentations, yet are not ready to delve into programming. Here we highlight three presentation programs.

Lotus Freelance   Freelance has been the most popular of these programs. It is very easy to use. A Freelance presentation can be played back on a DOS machine that does not have Windows or Freelance on it by using a player program included with the product.

Freelance contains a variety of drawing tools. In addition, it supports the **import** of a variety of popular file formats. It also allows you to **export** the image of a slide in a variety of file formats, so it can serve as a simple drawing program. It provides a light-table feature that fills the screen with thumbnail sketches of the slides; you can easily drag slides around, thus graphically rearranging your presentation. Another nice feature is a "slide build" feature, which can present bulleted items one at a time, so the audience can concentrate on the message of the speaker instead of reading ahead. Previously displayed bulleted items on the same slide are dimmed. Styles (a template for a slide) maintain visual consistency from slide to slide, and common elements, such as a graphic or a title, can be made to appear on every page by adding it to the master page.

The person creating the presentation may place clickable button images on the screen, or screen objects (such as the text) can be made sensitive to mouse clicks. These objects can then respond to mouse clicks by the presenter, causing the presentation to branch to another screen, launch another program, or play a multimedia file.

On-screen presentations take advantage of the full screen; in other words, there are no distracting Windows borders and menus to detract from your material. A nice feature is "digital chalk," whereby you can use the mouse to draw on the screen during the presentation.

Asymetrix Compel   This is a presentation program that specializes in multimedia. It lacks some features of the other products, such as an outliner and a spelling checker. It shares many features with Freelance, such as the light table, selection of drawing tools, import and export capability, and slide build feature. A Compel presentation may be delivered on a computer that does not contain Compel through the use of a player program shipped with the product. However, unlike Freelance, the computer must have Windows installed.

Multimedia is Compel's forte. All objects can be animated, so, for example, bulleted items can move into place from the bottom of the screen. Animations can only proceed along a straight line, but the speed of the animation can be controlled, and the object can change size progressively during the presentation. An object may be made visible or invisible before and/or after the animation. Any object can trigger other events when it is clicked, or when its animation starts or stops. You can create, with Compel, a set of clips (subsets of the multimedia file, e.g., part of a song) from multimedia files. These clips can be given a name and associated with one or multiple objects so that the clip will play in response to clicking any of the objects. Multimedia play can also be triggered by a new slide being shown or by a slide exiting the screen.

There is no "digital chalk" feature. However, Compel has rich text function. You can create a scrollable text box on a slide, and scroll it with the mouse when the slide is shown during a presentation. Any character or set of characters (such as one or more words) can be defined as a **hotword,** which can be assigned behavior just like other objects.

There is a checking feature that makes sure all the files needed (the presentation file plus all the invoked multimedia and other files) are available. Compel can also package all the needed files into a single presentation file, which is a convenience when you need to copy your presentation to another computer.

Microsoft PowerPoint   This was the first presentation graphics program for the Windows platform. It is as easy to use as products such as Freelance. It is a full-featured product, which includes the slide sorter feature, slide transition effects, slide build capabilities, an outliner, spelling checker, and digital chalk. Microsoft PowerPoint includes templates and styles for many presentations. It also includes a unique color selector, which is handy if you do not wish to stick with the color schemes of a template. You choose a background color, and PowerPoint presents you with a choice of foreground colors that work well with the background color you chose. Once you choose the foreground color, PowerPoint presents you with several sets of highlight colors that complement your choice of foreground and background colors.

PowerPoint includes a viewer program that you can include with your PowerPoint presentation. This viewer, which you can distribute royalty-free, allows one to play a PowerPoint presentation on a machine that does not have PowerPoint installed.

Choosing a Program   If your needs are limited to presentations, these packages will serve you well. They might also provide a means for you to get your feet wet with multimedia before you move on to an authoring program. Freelance has some nice features such as spelling checker, "digital chalk," and the ability to run the presentation on a DOS machine. Compel has more multimedia prowess and simple animation. Compel can also be used to create simple hypermedia interactive programs, such as tutorials, for student use. Students interact with the program by clicking on text and objects with the mouse pointer. Other programs with unique capabilities include Macromedia Action and Gold Disk's Astound.

AUTHORING PROGRAMS

If you are interested in multimedia and want to create finished applications, you will probably be looking at authoring programs. This category of programs allows you to create complete hypermedia applications, with rich interactivity features and the ability to capture, analyze, and store student input.

What is unique about these programs is that they allow you to implement the user interface elements of the program visually. For example, if you want some text at a location on the screen, you just select a text tool, create a container for the text, and type it in. Any formatting can be done on the spot. Don't like where the text is? Just drag it and move it. This is very different from the way it was done with traditional programming languages, where if you wanted text on the screen, you had to write program statements to put it there. And you could forget about fonts; the only choice was monospaced text in a single point size.

Authorware Professional This program is available in both Windows and Macintosh versions, and is highly respected. You create a program in Authorware by drawing a flowchart that describes the function of the program. A sample flowchart is shown in Figure 61. Different types of blocks are available for inclusion in the flowchart, such as blocks that display a screen, that erase objects, that animate them, that obtain user input, that branch based on the user's input, and so on. There is no programming language to learn. Drawing and text tools are included that enable you to create the screens.

User interactivity is very rich. The program can accept input in the form of text, keystroke responses, menu selections, button clicks, and even the dragging and dropping of an object at specified screen positions. A number of functions help you to analyze text input. For example, you might provide a number of acceptable responses for a question, so the program will accept various answers as correct. Your Authorware program's branching logic can require that an answer be given within a time limit; perhaps at the end of the time limit you would like to provide the student with a hint. Another option is to allow the student a number of retries for a specific entry before the program offers a hint.

The program has excellent animation capabilities. Objects can be animated along straight lines or author-defined paths. The speed of the animation is program controlled, and the position of an object can be made to correspond to the value of variables. Another Authorware feature is that the program keeps track of a number of parameters, such as how long the program has been running, how

FIGURE 61 *An Authorware Flowchart*

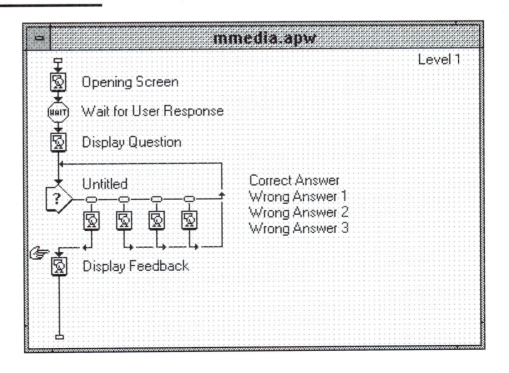

long it has taken the student to respond to a question, how many tries were required to answer a question correctly, and so on (there are more than 100 such variables). The program can make decisions based on the values of these variables, and their values can be stored in a file.

Because Authorware predates the availability of MCI (see page 94), Authorware implements its own multimedia function. However, the latest version of the product also supports multimedia through the MCI interface.

Authorware is not as well suited for hypertext applications, because the multiplicity of navigation paths through such a program cannot easily be represented by a flowchart, the fundamental paradigm of Authorware.

Multimedia ToolBook Multimedia ToolBook is a general-purpose Windows application-development tool. It is especially adept at hypermedia. ToolBook follows a book metaphor. A ToolBook application is called a book, and each of its screens is a page. Multiple pages can share a background, on which common elements can be placed and will show on each page.

You write a ToolBook program by creating pages and placing the program's user-interface objects (buttons, menus, text, graphics) on them. Each object can be programmed, using ToolBook's OpenScript language, to respond to events that concern them. For example, you can program a button to respond to a mouse click by causing the program to jump to a different page. You have total control over all of the properties of all the objects, both at design time and when the program is running. You can change an object's appearance, its size, its location, or whether or not it is visible. Any text, or portion of text, can also be an object, so that it too can respond to external events. For example, we might make a technical word into an object, and program it to make visible a box containing a definition of the word when the user clicks on the word. Thus the definition is available only when and if needed, and does not take up screen space when it is not needed.

You can have as many objects on a page as you want, and each can be programmed to cause the program to branch to any other page. Thus ToolBook is well suited for programs containing a lot of linked information, where the student can follow the information in a variety of ways. Interaction between the program and the user can occur in an almost infinite number of ways. An object can respond to mouse clicks or to having the mouse pointer move over it. Objects can respond to individual keystrokes, or the program can be designed to accept a large amount of text before processing the text.

Multimedia ToolBook has good support for all multimedia devices. A number of predesigned objects are provided that control multimedia. These objects can be copied into your own ToolBook programs and customized. Some of these objects take the form of buttons, or sets of buttons, resembling the controls on a tape recorder, so they present the user with a familiar interface.

The product has a rich set of features to ease the development task, such as a clip-art collection, drawing tools, a script recorder (that can record your actions and generate code that duplicates the actions), a good number of sample programs, a competent debugger, and so on. We will see a lot more of Multimedia ToolBook in the next chapter.

A limitation of ToolBook, when compared with Authorware, is that ToolBook does not provide built-in courseware management functions such as scoring, answer judging, and so on. While all of these can be built within a ToolBook application using the OpenScript language, this represents extra work for the ToolBook application developer. However, in late 1994 Asymetrix released Multimedia ToolBook 3.0 CBT Edition; this version of ToolBook includes a wealth of built-in courseware management functions, including automatic scoring, stu-

FIGURE 62   *A Visual Basic Form with Controls*

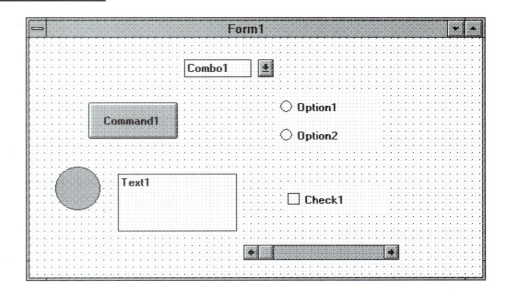

dent data tracking and logging, reporting, bookmarking, and so on. In addition, the product provides a number of courseware templates to ease the task of the developer using ToolBook to create instructional materials.

Another Asymetrix product, the ToolBook Database Connection, provides ToolBook with access to many popular **database** formats.

Visual Basic   Visual Basic is a combination of an interactive design tool and a powerful programming language. To write a program, you design your program's screens (called *forms* in Visual Basic) interactively by placing user-interface objects (called *controls*) and customizing them. Figure 62 shows some interface controls on a Visual Basic form (window), while Figure 63 shows the Visual Basic Properties window, which is where you set the values of the properties for a Visual Basic control. Just as in ToolBook, these controls can be programmed to

FIGURE 63   *Properties Window in Visual Basic*

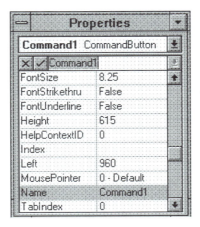

respond to events that concern them. The code you write becomes a part of the control, so it is easy to find and modify.

The BASIC language of Visual Basic is a far cry from the BASIC of yesteryear. It contains most of the features of more advanced languages. The GO TO statements and line numbers of traditional BASIC programs are nowhere to be found.

There are two versions of the program: Standard and Professional. The Professional edition adds features such as 3D controls, a graphing control, and multimedia capabilities. The multimedia control provides basic multimedia function, allowing you to play multimedia files, pause them, and so on. However, if you want more sophisticated control, such as playing only a portion of a file, you will have to write programs to do it.

One of the best features of Visual Basic is its ability to incorporate controls written by third parties. There are many of those on the market. For example, if you want a sophisticated editing control in your program, it is available. If you need **TCP/IP** support, there is a control that provides that. How about better multimedia controls? Yes, those are available too. The effect is to make the product richer than it already is.

Visual Basic has excellent database capabilities, as it includes the Microsoft Access 1.1 database engine. The Standard edition can access databases in many popular formats. Data can be read and updated, and new records written, even with databases composed of multiple tables. Although the Standard edition cannot create a new database, the product includes a separate tool that provides this function. The Professional edition is capable of creating databases, and of performing just about any sophisticated database task that you might require. Included with the Professional edition is Crystal Reports for Windows, so that you can interactively create database reports and run them from within a Visual Basic program.

Visual Basic has a number of tools to help you create a program. The code editor color codes your program's statements, so the program is easy to analyze and follow. There is an excellent debugger. Each form is stored in its own file, so it can be used in a number of programs.

Visual Basic combines interactive, visual program design with the full power of a modern BASIC language.

Other Programs   The three software packages described above are well suited to those getting started in education software development. But be aware that there is a great variety of authoring and development tools on the market. IconAuthor, like Authorware, is based on the flowchart metaphor. CA Realizer is another visual BASIC development tool. For the sophisticated programmer there are many C and C++ systems to choose from. Those interested in a purely object-oriented language can choose from several varieties of Smalltalk. Pascal aficionados will find their needs met by Turbo Pascal for Windows and Delphi. Delphi, by Borland, is similar in concept to Visual Basic. However, it is richer, and is object oriented.

Choosing a Program   If you plan to choose from among the three programs described above, you might feel that the descriptions provided are not adequate to enable you to make a decision. For this reason, a comparative table (Table 7) is provided.

If your application is fairly linear and can easily be described by a flowchart, then Authorware might be a good choice. If you need a lot of hypertext, or want applications with many screens with lots of text or common elements, or if you need to convey a lot of information, then ToolBook is a good choice. If your program resembles a typical computer program, with a lot of calculations or process-

TABLE 7 ▌ *Comparison: Authorware, Multimedia ToolBook, Visual Basic*

Feature	Authorware	Multimedia ToolBook	Visual Basic Professional
Ease of learning	Easiest	Moderate	More difficult
Tracking user progress	Built in	Can be programmed into application	Can be programmed into application
Multimedia capability	Good	Excellent	Limited, but can be programmed into application
Hypertext	Very limited	Excellent	Does not lend itself well to hypertext
Programming language	None: the flowchart is the language	Yes: OpenScript	Yes: BASIC
Completeness of language	Not applicable	Missing: data structures	Complete
Database capability	None	Limited	Excellent
File handling	Very limited	Limited	Excellent
Link to Windows **DLLs**	Yes	Yes	Yes
Third-party extension products available	No	No	Yes
Cost	Expensive	Less expensive	Least expensive
License restrictions	Some	None	None
Color support	256	16.7 million	256

ing, or if you need special features (such as TCP/IP capability), then choose Visual Basic.

MOVIE-EDITING PROGRAMS

This is a new field and thus is not as crowded as some of the others covered above. Fortunately, this does not mean that there is a lack of good programs.

Adobe Premiere   This program was first available on the Macintosh, but a Video for Windows version was shipped in September of 1993. Because Adobe Premiere was discussed in the section "Working with Digital Video" on page 68, it will not be covered here.

Choosing a Program   If you create Video for Windows movies, you should look at Adobe Premiere. If you create hardware motion video movies using the ActionMedia II card, the D/Vision program (which runs on DOS systems) provides function similar to Premiere's. If you need good video editing capability, but your needs do not demand the high power of Adobe Premiere, you might be very pleased with Digital Video Producer, a similar product which is a part of the Multimedia ToolBook 3.0 product.

Drivers

Drivers are special programs that make the connection between the software and specialized hardware devices. Drivers are needed for such things as an audio card, a video card, MIDI, and so on. Some software systems also require playback drivers. To play an animation file or a Video for Windows file you will need the proper driver.

Normally you do not have to worry about them. When you purchase a card, it will come with a diskette containing the driver, along with installation instructions. If a program comes with animation or software motion video files that it plays, the program will, as a rule, include the proper drivers and install them as a part of the program's installation procedure.

However, drivers sometimes can cause problems. A new program might require a newer level of a driver than the one you have installed. There might be an interrupt conflict between a new card and the rest of the system. You should be aware that these problems do occur, and they will generally manifest themselves after you install some new hardware or software. If you cannot solve the problem on your own, consult with the developers of the new item you installed. Chances are they have seen that same problem in other systems and have a solution already worked out.

12

Multimedia ToolBook

In the previous chapter we surveyed the field of multimedia software. Although you now have a feel for what is available, we did not cover any one program in sufficient depth that you have an understanding of how it works. In this chapter we will look in more detail at one multimedia program: Multimedia ToolBook. Why this particular program? For several reasons. Multimedia ToolBook 3.0 includes most of the tools to perform any kind of multimedia work. Multimedia ToolBook has enjoyed success as a tool for developing higher education instructional materials. It has been used to author several educational products, such as Microsoft's Multimedia Beethoven, The Enduring Vision (an American History textbook on CD-ROM), and Interactive Calculus (also on CD-ROM), among others.

In order to understand and appreciate the power of this tool, however, we must first recap the evolution of computer programming languages.

Programming

Computers are and have always been general-purpose tools. What makes a chess program or a word processor different from a program such as Microsoft Flight Simulator is the software the computer runs. The program is a collection of instructions that tell the computer in detail what to do. A program to print someone's paycheck contains instructions to look up the person's salary or wages from a file, compute taxes and deductions, print out the proper check, and deduct the amount from the firm's bank account. What makes computers seem intelligent (to some degree) is that computers are able to test data and make decisions on what to do next based on the results of the test. The program at the bank that processes your checking account will be given the amount of a check submitted for payment, and your account balance. It will test the balance to see if it is greater than or equal to the amount of the check. If it is, the program proceeds to deduct the check amount from your account and to credit it to another account. If the check amount exceeds your balance, the program skips (branches) to another set of instructions which will reject the check and perhaps write a nasty letter to you.

Because computers are binary, in other words, work with 1's and 0's, ultimately every computer instruction boils down to a sequence of 1's and 0's. Programmers writing programs for the first computers had to write each computer instruction as a sequence of 1's and 0's, and then write each such sequence into the computer by using a set of switches on the computer's control panel. This was very slow, tedious, and error-prone work.

Errors in computer programs, and glitches or design errors in the hardware, are called *bugs*. The process of tracking down and correcting these errors is called *debugging*. We owe the words "bug" and "debugging" to Admiral Grace Murray Hopper (1906–1992), a pioneer in the computer programming field. She was working on the Harvard Mark II computer when it malfunctioned. She traced the problem to a moth (a bug) that was caught in the contacts of a relay in the computer.

Computer programmers soon tired of writing 1's and 0's. They developed a symbolic language with which the programmer could represent each instruction with a mnemonic that was much easier for the programmer to remember. An instruction that moved data from memory to the computer's registers could be written as LOAD. An add instruction would be coded as ADD, and so on. A program in the computer, called an *assembler*, translated each mnemonic in the program to the corresponding sequence of 1's and 0's.

Later on, higher-level languages were developed. In these languages, one instruction might correspond to several machine instructions. For example, to add two numbers might take three or four machine or assembler instructions to load the data into the computer registers, perform the addition, and store the result back in memory. In a high-level language, this could be expressed in understandable notation, such as

$$A = B + 5$$

The first of the high-level computer languages, which are still in use today, were FORTRAN (for FORmula TRANslator), a language devised for engineering and scientific computing by John Backus of IBM, and COBOL (for COmmon Business Oriented Language), which originated with Grace Hopper. Many other languages followed, such as Algol (for Algorithmic language), APL (A Programming Language), PL/1 (Programming Language 1), Pascal (after Blaise Pascal, a mathematician), ADA (named for Pascal's wife), and C (it followed the A and B languages at Bell Labs), along with more obscure languages such as Forth, SNOBOL, Prolog, and so on.

These languages ran on **mainframe** computers, and most of them were batch oriented. This means that you loaded the program onto the computer, probably by having the computer read in a deck of punched cards, each card containing one computer instruction; started the program; and the program would run unattended until it completed. In the 1960s, two professors from Dartmouth College came up with a simple, easy-to-learn language they called BASIC, for Beginners All-purpose Symbolic Instruction Code. Besides being simple, it was also, in a way, interactive. You interacted with the BASIC interpreter in writing a BASIC program. You could type in a BASIC statement and have the computer run it right away. Thus BASIC was well suited as a first computer language. BASIC ran on mainframes, but you would interact with it through a teletype terminal connected to the computer.

The early 1980s saw the coming of age of the personal computer. Perhaps the first successful personal computer was the Apple II. The IBM personal computer, with its open architecture, spawned an industry of "IBM-compatible" computers (clones). The IBM PCs and clones came with the BASIC language interpreter, to enable users to write their own programs. BASIC was a good language for personal computers because it was simple (many purchasers of personal computers had never programmed before), small enough to fit in the tiny memories of the original PCs (the base memory on the original IBM PC was 16K), and its interactive nature was a good match for the PC's interactive style.

Another revolution came in 1984 when Apple introduced the first Macintosh computer. The Macintosh sported a small screen and had no color capability; it was compelling nonetheless, because it provided a graphical user interface (GUI), so users could accomplish work by clicking on icons and selecting commands from menus instead of typing them. The graphical nature of the Macintosh made it possible for people to easily incorporate graphics into their text documents. This is because the text itself was graphical in nature, whereas on the PC, text and graphics were two distinct entities. Thus the Macintosh became the preferred system for desktop publishing and graphical applications.

Eventually Windows was able to provide a graphical user interface for the PC. The problem for both the Macintosh and Windows was that graphical programs are much harder to write. Users expect programs for the Macintosh and Windows to provide menus, dialog boxes, buttons, scrollbars, and other graphical controls. They expect to be able to render text in a variety of fonts and colors. They also expect the programs to provide a high degree of interactivity.

The programming languages then available were inadequate for this task. Most languages had no graphics capability. They could write text to the screen and accept text the user typed in. However, they could not deal with fonts. BASIC provided some graphics capability, but was still laborious to use as a graphics authoring language. For example, to draw a rectangle on the screen required the programmer to write a statement providing the pixel coordinates of the corners of the rectangle.

The breakthrough came with the introduction of HyperCard by Apple. This program introduced the concept of hypertext, and made it possible for many, especially in higher education, to write applications for the Macintosh with much less effort and programming skill than was previously required. A few years later, with the introduction of Windows 3.0, ToolBook was made available for the Windows platform, providing capabilities similar to those provided by HyperCard. We shall now take a closer look at ToolBook.

ToolBook Concepts

ToolBook uses the book metaphor. A ToolBook program, then, is known as a *book*. The individual screens or windows of the program are called *pages*.

Think of an encyclopedia program. As implemented in ToolBook, each entry would be contained on a page, which would contain text, graphics, and perhaps some multimedia content, such as a sound file or a movie. In the text there would be cross references to other entries in the encyclopedia, indicated perhaps by the text being in a different color from normal text. If the user were to click on such a word, ToolBook would bring up the proper page for display. This ability to move through the program in a nonsequential manner is an important characteristic, as it allows the user to determine his or her own progression through the material.

The Fundamental Elements of ToolBook

There are four elements that make up a ToolBook program, or book: (1) objects, (2) properties, (3) messages, and (4) script. We will discuss each of these elements individually.

OBJECTS

One of the most important recent developments in programming is object-oriented programming (OOP), where the program is made up of objects. Object-oriented programming is important because objects are a more faithful representation of the real world than are traditional programming structures. Object-oriented programs are easier to maintain because the objects are self-contained, so changes in an object's implementation will not affect the rest of the program. In addition, objects are reusable in other programs, reducing the time required to write them. Despite these advantages, object-oriented languages are often harder to learn than traditional programming languages.

ToolBook takes advantage of some of the object-oriented concepts without being burdened by the complexities of object-oriented programming. This is best illustrated by example. Let's contrast drawing a rectangle in BASIC and in ToolBook. In traditional BASIC you write a programming statement (the LINE statement) to cause the program to draw the rectangle. You can also use a keyword in the statement to specify that the rectangle be filled. You then run the program (or at least the statement) to see its effects. If the rectangle is not quite in the right place or of the right size, you edit the statement to change the coordinates and repeat the process until you get it right. Once the rectangle is drawn, the program has no knowledge of it. The rectangle is simply a set of pixels on the screen in a color that is different from the color of its surroundings.

To erase the rectangle you must change the color of the pixels composing the rectangle. You must remember the coordinates of the rectangle, as the program has no knowledge of its existence. Moving the rectangle consists of taking the block of pixels and copying their color values to another location on the screen. Then you must go back and restore the pixels where the rectangle "was" by changing their color to the color they had before the rectangle was drawn.

Changing the size of the rectangle is also a complex undertaking. These tasks can be even more complex if the picture on the screen is not rectangular.

Everything in ToolBook is an object. To create a rectangle, you click on the rectangle tool on the Tool Palette (Figure 64), and draw the rectangle by dragging the mouse pointer on the screen. To change the position of the drawn rectangle you simply drag it to a new position. To change its size, drag a handle (the small rectangles surrounding the object) until the rectangle is the desired size. Programs like ToolBook also let you color and add a pattern to the inside of an object (Figure 65) with a couple of mouse clicks. One difference, then, between ToolBook and BASIC is that in ToolBook you create your screens interactively, instead of by writing programming statements.

Once you have created the rectangle, ToolBook treats it as an object. Instead of manipulating pixels, you interact with the object itself. You can give the object a name and refer to it by name in your ToolBook program. Say we name the rectangle "Happy"; we could then write statements such as

Move rectangle Happy to 100,100

or

Hide rectangle Happy

Thus the rectangle is not simply a collection of pixels on the screen. It is an entity that ToolBook knows about and can treat as a unit. When you move the object, ToolBook handles the painting of the screen region the object has vacated. This applies to all objects, regardless of their shape. Objects in ToolBook include graphics, buttons, fields (objects that hold text), pages, books, and so on.

FIGURE 64   *Multimedia ToolBook Screen*

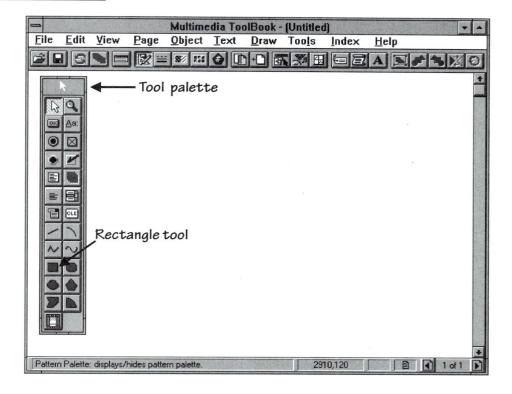

Creating complex user interfaces for programs is very hard work. ToolBook makes the task much easier because you can design your screens interactively and because it makes available a rich selection of objects from which to build your programs.

This applies to text as well. Unlike BASIC, where text is limited to character positions on the screen (normally 80 characters per line, 25 lines per screen), in ToolBook text can be placed anywhere. To place text you type it in place exactly where you want it. Text can be rendered in any available font and color, and you can mix fonts and colors within the text.

A ToolBook book is composed of one or more pages. What you see on the screen is a combination of a page and a background (both of which are also ToolBook objects). The background is displayed underneath the page; it shows through anywhere on the page where there is an empty space. In other words, the page itself is always transparent, although the objects you place on the page (buttons, fields, graphics) normally are not. Backgrounds can be shared by many

FIGURE 65   *Rectangle Object in ToolBook, filled with happy-face patterns*

pages. Thus if you have an object that you want to have appear on every page, such as a button that when clicked, turns the page, you can put the object once on the background instead of having to put a copy of it on each page.

PROPERTIES

All objects in ToolBook have properties, which determine what the objects look like and how they behave. Properties include the position of the object (the position property), the size of the object (the size and bounds properties), and its colors (the strokeColor and fillColor properties), among others.

You modify the look of objects in a ToolBook program by changing the value of their properties. To move an object you change its position property, which can also be done by using the Move command. To change the text in a field, you modify the field's text property. If the user types text in a field, you can examine the text by retrieving the field's text property.

A very useful property is called *visible.* If it is set to True, the object can be seen on the screen. If it is set to False, the object is still present, but it cannot be seen, and anything behind it can be seen as if the object were not there. Changing this property is an easy way for you to make objects appear and disappear from the screen.

An object contains all of its properties. When an object is created, ToolBook assigns it a set of default properties. Other properties, such as position and size, are determined by where the object is drawn. From then on you are free to assign values to any of the object's properties.

It is easy to see how properties can affect the appearance of an object. Changing the position or one of the color properties of an object will obviously change its look on the screen. However, properties can also affect how an object behaves. If you set a button's "enabled" property to False, you will change its appearance (its caption, or label, is dimmed) and also its behavior (the button will not respond to mouse clicks). Setting a fields' "activate scripts" property to True will prevent the user from being able to type in it.

Although ToolBook provides its objects with a comprehensive set of properties, the designers of ToolBook cannot anticipate the needs of all authors. Thus ToolBook provides you with the capability to define and implement additional properties for ToolBook objects. For example, you might define a book property called "bookmark" which keeps track of the last page the user visited. Thus the next time the user opens it, your program can start up at the page where the user left off. ToolBook does not include such a property, but you can easily define and implement it.

MESSAGES

One of the aims of ToolBook is to allow you to create interactive programs. We have already seen that ToolBook is very interactive with the programmer, allowing you to build the screens, or pages of the book, by drawing and writing directly on the screen. ToolBook also enables you to create programs that are highly interactive with the ultimate user.

The mechanism to accomplish this is the message, which is used to notify an object of some action by the user. Any user action, such as moving the mouse pointer, clicking a mouse button, selecting a menu item, typing at the keyboard, and so on, causes ToolBook to generate messages that it sends to the appropriate object. Let's clarify this with an example:

Suppose you place a button on the page, whose function is to turn to the next page when the user clicks it. It is the responsibility of the button, when clicked, to cause ToolBook to turn the page (we will see how this is done shortly). When the user clicks on the button, ToolBook sends the button a message, indicating that the button has been clicked. Think of ToolBook as a lookout on behalf of your program. When anything of interest to the program happens, ToolBook lets the program know.

When an object receives a message, it can act on it or it can ignore it. If it acts on it, that is the end of the message. However, if the object does not respond to the message, ToolBook forwards the message to the next object in the hierarchy. Briefly, pages are higher in the hierarchy than regular objects such as buttons, graphics, and fields, backgrounds are higher than pages, and books are higher than backgrounds (Figure 66).

For example, suppose the cursor is in a field and the user starts typing. With each keystroke the user types, ToolBook sends a number of messages to the field object. If the field ignores the messages (we will see how that is done in the next section), the messages travel up the object hierarchy until they get to ToolBook itself, assuming no other object in the hierarchy handles the message. When ToolBook receives the messages, it performs the necessary steps to place the typed characters in the field. This means that your program can examine the characters, and if they are not what the program expects, the program can prevent the message from reaching ToolBook, so that no characters

FIGURE 66   *Simplified ToolBook Message Hierarchy*

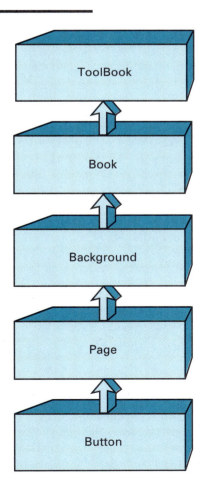

appear in the field. You can think of ToolBook as having two parts: The part that we have called the "lookout," which monitors events and generates and directs messages, and the other part, at the top of the hierarchy, which receives and processes messages that are not handled by the objects in the hierarchy.

ToolBook gives you great flexibility in creating the function of your user interface by providing a rich selection of messages to let your program know exactly what the user is doing. Let's look at the messages that can be generated by the user's actions with the mouse:

mouseEnter: This message is sent to an object when the mouse pointer crosses over the object's boundary. If the object is a graphic object, it does not matter how irregular its border is; ToolBook will send the mouseEnter message exactly when the mouse pointer crosses the border. You might have a map of a state on the screen. As soon as the user moves the mouse pointer into the state, the state graphic object receives the message, and might take some action, such as displaying the name of the state. As another example, consider that some newer programs, such as Microsoft Word for Windows 6.0, CorelFLOW 2.0, and Multimedia ToolBook 3.0, will display a short message indicating the function of a button on a tool bar when the mouse pointer is inside the button. You can implement a similar function in your program by using the mouseEnter message.

mouseLeave: Many ToolBook messages come in pairs. An object receiving a mouseEnter message will receive a mouseLeave message when the mouse pointer again crosses the object's boundary.

buttonClick: An object receives this message when the user clicks the left mouse button (presses it and releases it) when the mouse pointer is within the object's boundaries.

buttonDown: This message is sent to the object when the user presses the left mouse button down while the mouse pointer is within the object's boundaries.

buttonUp: This message is sent to the object when the user releases the left mouse button. It does not matter where the mouse pointer is when the button is released. If an object received a buttonDown message, it will receive a buttonUp message when the button is released.

buttonStillDown: If the user keeps the left or the right mouse button pressed, ToolBook will issue successive buttonStillDown messages until the button is released. This message is useful in some drag operations.

buttonDoubleClick: This message is sent to the object when the user double clicks the left mouse button while the mouse pointer is within the object's boundaries.

rightButtonDown: This message is the same as buttonDown, but for the right mouse button.

rightButtonUp: This message is the same as buttonUp, but for the right mouse button.

rightButtonDoubleClick: This message is the same as buttonDoubleClick, but for the right mouse button.

You can see that you have a lot of flexibility in implementing your program's interaction with the user thanks to the variety of messages that ToolBook sends an object when it is acted upon by the mouse. The above list does not exhaust the number of messages generated by the mouse.

There are many other messages that do not directly correspond to user's actions. Each time the user turns pages in a ToolBook book, a leavePage message is sent to the page the user is leaving, while an enterPage message is sent to the page the user is turning to. These messages can be used to initialize the contents of a page (set the contents to an initial, undisturbed state), perhaps to move an animation back to its starting point before the user sees the page again. The messages enterBook and leaveBook are sent to the book when it is opened and when it is closed. You might use the enterBook message to perform initialization when the book (program) starts up, and the leaveBook message to save information before the book shuts down.

SCRIPT

Although ToolBook provides a variety of objects, properties, and messages, you will soon run up against a wall, meaning that you will want to implement something that ToolBook cannot do. This is a problem with many authoring programs that do not provide a programming language. Although such authoring systems are easier to learn, there is a significant limit on what they can do. ToolBook provides its own programming language, called *OpenScript,* which allows you to expand the capabilities of the product. As a matter of fact, very few ToolBook applications are created which do not contain programming in OpenScript.

The language is termed a "script" as opposed to a programming language because it is a little easier to learn and use than a regular programming language. It is more readable as well. It is "English-like." Remember, though, that you are still dealing with a computer, no matter how English-like the language might seem. While humans can easily recognize spelling errors and infer what the author meant, ToolBook and most other programs cannot.

Each ToolBook object contains its own script. The script of an object is one of the properties of the object. The script of an object is what gives an object its function. An object's script consists of a number (zero or more) of "handlers." A handler is a unit of code that responds to a message received by the object.

Let's return to the button that should turn the page when clicked. If you write no script for the button, when the program is running and the user clicks the button, the button will appear to move in and out. This is the built-in functionality of the button object. However, nothing else will happen, because the button does not "know" what else to do when it is clicked. The script that you write tells the button what to do. The script for this button would be:

```
to handle buttonClick
    go to next page
end buttonClick
```

Now, the user clicks the button. The button receives three messages: buttonDown, buttonClick, and buttonUp (let's ignore the mouseEnter message). When each message is received by an object, ToolBook checks for a handler in the object with the same name as the message. If such a handler is not found, the message moves up the hierarchy. If such a handler is found, the code in the handler is run.

The button in our example receives the three messages. ToolBook determines that there are no buttonDown and buttonUp handlers in the button's script, so the messages travel up the hierarchy and nothing happens. When the buttonClick message is received, ToolBook determines that there is a handler for this message in the object's script, so it runs the script, which causes ToolBook to turn to the next page in the book. If the button has multiple handlers (multiple pieces of code that start with the "to handle..." statement), then it will respond to multiple messages, and thus have richer function.

Multimedia in Multimedia ToolBook

Multimedia ToolBook supports digital audio, MIDI, CD-ROM audio, animations, videodisc, videotape, digital video, and images (bitmaps and Photo CDs). Each piece of multimedia is itself an object. Properties for such objects include the file name (for file-based multimedia such as digital video, digital audio, and MIDI), the starting and ending positions of the clip, and so on.

Changing the properties of a multimedia object will affect the object by, for example, changing the name of the file which is the source of the clip or changing the starting and ending positions of the clip within the multimedia object. Multimedia objects (clips) are played by executing the ToolBook OpenScript mmPlay statement. There are several other multimedia commands that operate on clips, letting you pause, stop, rewind, go to a specific place in the clip, and so on. You create the clips interactively with tools built into ToolBook.

Clips that provide images, such as animations, stills, and digital video, need another ToolBook object, the *stage,* which determines where on the screen the image(s) will show. The stage object also allows you to define any cropping or stretching of the visual media, as well as transition effects to introduce and end the clip.

A digital video player is built in, so you can add digital video to your program without writing a single line of code. Multimedia ToolBook provides a simple audio editor that allows you to record and edit audio for your ToolBook books. It also contains Digital Video Producer, a program that enables you to easily capture and edit digital movies. Multimedia ToolBook comes with all the tools you need for creating a multimedia Windows program.

Interactivity

We have already seen many ways in which ToolBook provides interactivity, both for the programmer using ToolBook to write an application and for the user using the program written in ToolBook. Creating and placing a graphic in ToolBook is trivial compared with doing it in a BASIC program. This interactivity is important, as it makes the task of the programmer easier.

There is another way in which ToolBook provides interaction with the programmer. ToolBook functions in two basic modes, Author and Reader. It is in Author mode that you create objects and write scripts. Reader mode is what the user of your program works with. When in Reader mode, your ToolBook program "runs." You can switch between Author and Reader modes by pressing the F3 key. Thus you can create a ToolBook object, write a handler for it, and immediately switch to Reader mode to run and test the program. This ease with which you can try out programs makes programming more efficient. Instead of writing a large program and then trying it (to find out, most of the time, that the program doesn't work) a ToolBook programmer can write a very few statements and try their function right away.

Case Study: A Multimedia Application in ToolBook

The purpose of this exercise is to see exactly what it takes to create a small multimedia program. Even if you do not have access to Multimedia ToolBook, this section will still be useful as an illustration of the process. This exercise is not to be considered a tutorial on Multimedia ToolBook. The program has so many capabilities that it would take a whole book to do it justice.

If you plan to implement the exercise, you will need a PC running Windows and Multimedia ToolBook, as well as a sound card, a microphone, a video capture card, and a video camera.

THE PROGRAM

Let's assume we want to create a foreign language program. On the first screen there will be a picture, some text (the sentence "Il était le matin du premier jour du printemps à Nice"), and a means for the user to hear the text spoken. On the second screen there will be a digital video, perhaps a video of Nice (you can fake it and take a video of your own neighborhood, as this is only an exercise). We will concentrate on the multimedia aspects of the program.

PREPARING THE MEDIA

Let's cover now how to prepare each of the media elements of our program.

The Audio Clip The first step is to record the audio. Multimedia ToolBook includes a simple audio capture and editing program called WaveEdit. When you bring up the program it will present a screen as shown in Figure 67.

With the microphone connected you can start the recording by clicking on the Rec button. If you do not have a microphone, you can load a sample audio file into WaveEdit to work with. Select Open from the File menu, and open a file in Microsoft Waveform format from the Multimedia ToolBook Samples directory.

The program will bring up the dialog box shown in Figure 68. Since what we are to record is the spoken word, we can settle for lower-quality audio. WaveEdit brings up a volume meter where you can calibrate the volume of your microphone. When you finish the recording, click the Stop button. Your screen might look somewhat like that in Figure 69.

At this point you can edit the recording by cutting and pasting parts of it. By moving the Zoom scrollbar you can zero in on a particular section of the

FIGURE 67 ▌ *The WaveEdit Application*

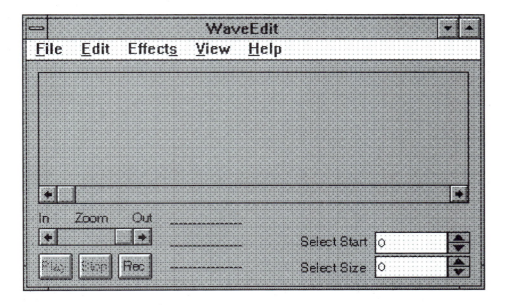

FIGURE 68   *WaveEdit—Selecting the Recording Format*

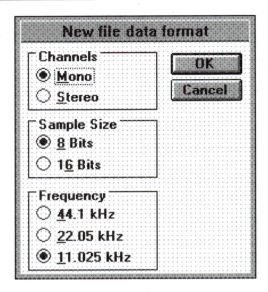

recording. By dragging the mouse cursor over a portion of the recording (with the left mouse button pressed), you can select any part of the recording for cutting and pasting.

If you are unhappy with the recording, just click New on the File menu to start again. Once the recording is edited to your satisfaction, select Save from the File menu. Save the recording with the name "voice" in Microsoft Waveform format.

FIGURE 69   *WaveEdit After the Recording*

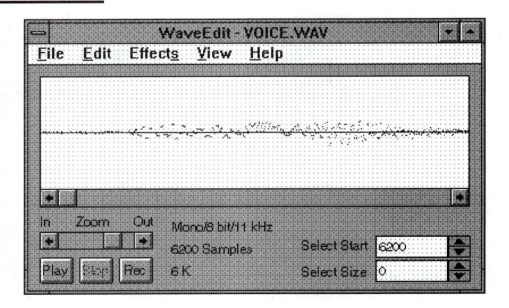

Is it really this simple? Yes and no. The mechanics of recording are simple. However, it takes much longer to master the art of recording. Professionals spend years learning how to set levels, mix sounds, and place microphones, in order to assure a quality recording.

The Video Clip   The mechanics of recording video are about as simple as those for recording audio, although the production of quality video takes a long apprenticeship. However, you can put together very good digital video with a little work.

Recording video consists of connecting a video source, be it a video camera, a VCR, or another source of NTSC (the North American television standard) to a video capture card. Depending on the card, you will need to capture the audio simultaneously using an audio card. Different cards come with different capture utilities, but they all work about the same. You click on the Record control to start the recording, and click on the Stop control to end it. Depending on the video capture card you are using, the video is compressed on the fly, or you will have to compress it later using a utility provided with the card.

Once the video is captured, you will likely need to edit it to cut out unwanted footage, to splice scenes together, and perhaps to enhance the quality of the video (color correction, filtering, etc.). For this you can use a program such as Adobe Premiere, or you can use the Asymetrix Digital Video Producer (see Figure 70), a program that is included with Multimedia ToolBook 3.0. Either program will take a little bit of learning before you can edit videos. If you do not have a video camera and capture card, you can practice with sample video files included in Multimedia ToolBook. When you are done, save your video as "nice.avi."

FIGURE 70 | *Asymetrix Digital Video Producer*

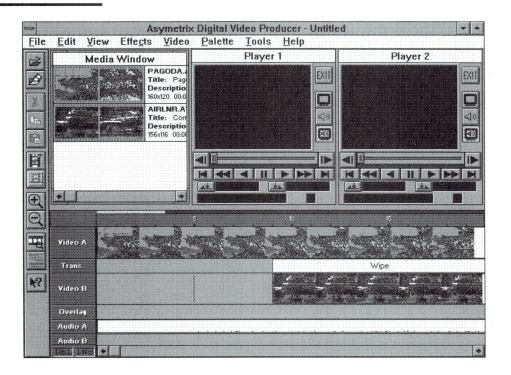

FIGURE 71 | *Multimedia ToolBook Tool Palette*

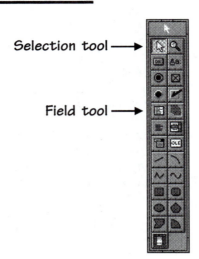

Selection tool →

Field tool →

WRITING THE PROGRAM

Bring up Multimedia ToolBook. It automatically creates a new book.

The Text   In ToolBook, text is contained in Field objects. Click on the Field tool on the ToolBook Tool Palette (see Figure 71).

The cursor will change into a "+" when it is positioned over the ToolBook page. Position the mouse cursor on a spot of the ToolBook page. Press and hold the left mouse button, and move the mouse down and to the right. As you do so, Toolbook draws the field. When you release the mouse button, the drawing of the field is complete; it looks just like a rectangle.

Click on the Selection tool. The cursor changes to an arrow. Click on the field to select it. The field becomes surrounded with 8 small squares called *handles*. Drag any handle to change the size of the field. Drag the field itself to position it on the page.

While the field is selected, double click on it. This puts you in text mode. The cursor changes to an I-beam when it is inside the field, and a blinking insertion point (a thin vertical line) indicates where the next character to be typed goes.

FIGURE 72 | *Field with French Text*

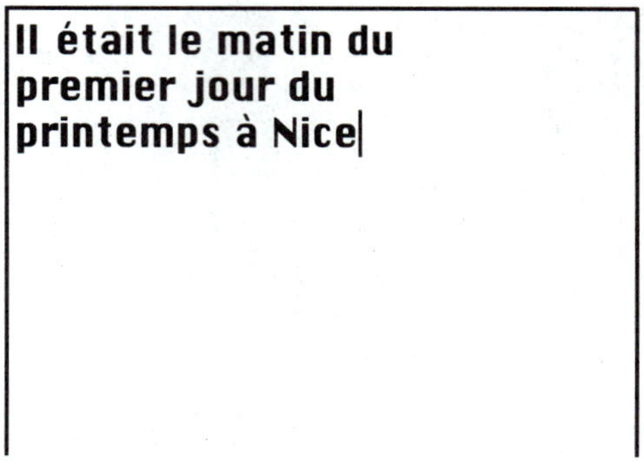

Il était le matin du
premier jour du
printemps à Nice|

FIGURE 73   *Popup Menu for a Field*

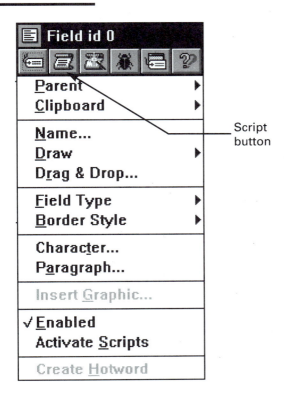

Script button

Type the text. Click on the Selection tool again, which takes you out of text mode. At this point your field should look like that in Figure 72.

Now format the text: Right click on the field. The popup menu shown in Figure 73 is displayed. Click on Character..., and a dialog box is displayed where you can select a font. After you dismiss the Character dialog box, right click on the field again, and select Paragraph... from the menu. Format the paragraph as Centered. The field pictured in Figure 74 was formatted with a 16-point New Times Roman Bold font.

The Audio and Video Objects We will design the program so that when the user clicks on the field, the audio that you recorded will play. The first step is to make the audio into a ToolBook object, called a *clip*. Open the Clip Manager (Figure 75) by selecting Clips... from the ToolBook Object menu.

Click on New, then on Sound (File) in the Choose Source Type dialog box. Locate the "voice.wav" file you created earlier, and select it. Click OK.

FIGURE 74   *Formatted Field*

Il était le matin du premier jour du printemps à Nice

FIGURE 75   ▌   *Multimedia ToolBook Clip Manager*

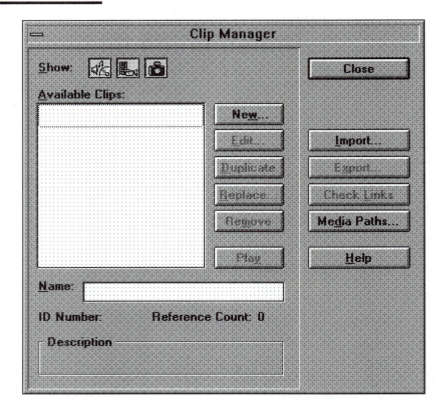

This brings up the Clip Editor, shown in Figure 76. With the clip editor it is possible to select just a portion of the file to be played. For example, you could create different clips from the same file so that when the user of the program clicks a word, the program plays only that portion of the file where that word is spoken. For this example, however, it is sufficient to play the file as a whole. Give the clip the name "voice", and click OK. You are returned to the Clip Manager.

Follow the same process to import the video file. In the Choose Source Type dialog box, choose Video (File). In the Clip Editor enter "video" for the name of the clip.

You are returned to the Clip Manager, which should now look like that in Figure 77. Notice that it now indicates the presence in the book of both clips.

Invoking the Audio   Now that the audio file has been made into a ToolBook clip, you can invoke it from within your program. Select the field again. Click the right-hand mouse button. Click on Activate Scripts on the popup menu—this enables the field object to execute code.

Right click on the field again. This time click on the Script button on the popup menu (see Figure 73). The Script Editor window comes up. Type the following script:

```
To handle buttonClick
mmplay clip "voice"
end
```

FIGURE 76   *Multimedia ToolBook Clip Editor*

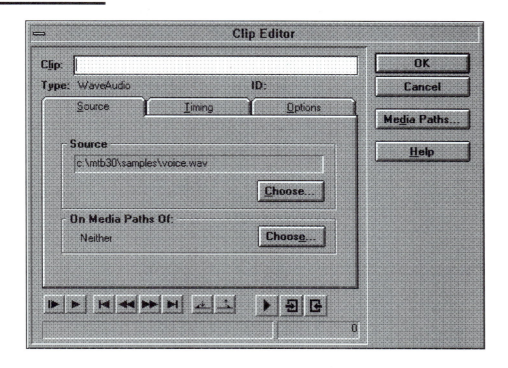

FIGURE 77   *Clip Manager with Both Clips*

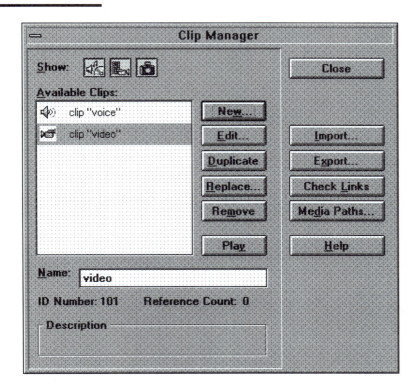

FIGURE 78   *Script for Field Object That Invokes the Audio Clip*

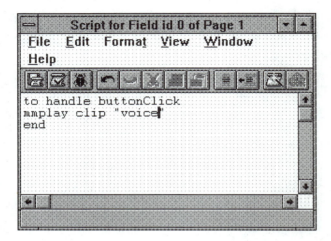

Your Clip Editor window should look like the one in Figure 78. This script (code, or program) does what might be suggested by reading the script: When the user clicks on the field with the mouse button, the field handles the click by playing the voice clip. Save the script by choosing Save Script, then Exit from the File menu in the Script Editor window.

You can test your work by putting ToolBook into Reader mode, by pressing the F3 key. The Tool Palette disappears, the menus are simplified, and your ToolBook application is running. Position the mouse cursor over the field and click. You should hear your clip. Press F3 again to return to Author mode.

Invoking the Video   We said that the video would be placed on a separate page. A new ToolBook book has only one page. To create a new one select New Page

FIGURE 79   *ToolBook Media Widget*

from ToolBook's Object menu. ToolBook creates a new page and turns to it. To switch back and forth between pages just select Next or Previous from the Page menu. This works in both Author and Reader modes. Although most programs provide navigation buttons to switch pages, we will leave that out of this program, as the purpose is to illustrate multimedia, not generic ToolBook authoring.

Adding the video is even easier than adding the audio because the ToolBook creators have done the work for you. Click on the Tools menu. From the pull-down menu, click on Media. Then click on Media Widgets on the cascading menu. ToolBook will present you with five options for adding the video to the page; the first one is shown in Figure 79. Choose this one by clicking on Copy to Book.

ToolBook will display the dialog box shown in Figure 80. Notice that ToolBook gives you several options on how to manage your video clip, such as playing it as soon as the user turns to this ToolBook page.

Click on Choose..., which brings up the Clip Manager. Click on the video clip to select it and click OK; after you close the various dialog boxes, ToolBook places the media widget into your ToolBook page, as shown in Figure 81.

Your program is now complete. Switch to Reader mode (F3), and click on the play button on the media widget. Your video should play. Go to the Page menu and select Previous. The first page should appear. Click on the field and the audio clip should play. You can now switch back to Author mode (F3 again), and choose Save from the File menu, to save your ToolBook program.

FIGURE 80   *Setting Options for the Media Widget*

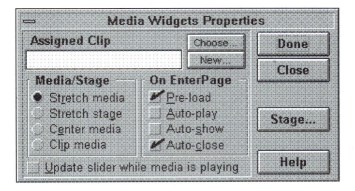

FIGURE 81   *Media Widget in Place*

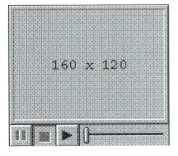

Conclusions

In this chapter we have looked in some detail at the architecture of a premier multimedia program, Multimedia ToolBook.

ToolBook incorporates some of the technology of object-oriented programming in order to make the programs easier to learn and use. In the future, we will continue to see the application of object-oriented concepts to tools of this nature, which will result in more powerful and yet easier-to-use programming tools.

We prepared a small sample program, involving the use of audio and video clips. If you followed the exercise in its entirety you will have noticed that preparing the media (the audio and the video files) was more work than incorporating the media into an application. Although we used a specific tool, Multimedia ToolBook, to create the program, the amount of work involved would be the same in other advanced multimedia authoring systems. The point is, given the advanced state of the authoring tools, creating multimedia applications is not difficult.

13

Developing a Multimedia Program

By now you have a good idea of what multimedia is and what the hardware and software options are. Perhaps you are now ready to start planning your multimedia application. This chapter contains some general guidelines intended to help you. It is written from the point of view of a professor or teacher approaching a multimedia project. However, the guidelines apply to anyone developing a multimedia program, be it educational software, a tool, or even a game.

Planning

Before you decide on your hardware needs and select your software tools, you need to plan. It is very important that you have a good idea of what you want your application to accomplish, and why. It is equally important to be very clear about how multimedia is going to contribute to the outcome. There should be a compelling reason to use multimedia. Otherwise, the extra complexity involved in multimedia will not be worth the effort. As we saw in the last chapter, creating multimedia materials is often a significant task compared with the implementation of the whole project.

Although programming can be fun, it is very time-consuming and there is a lot to learn along the way. Make sure you cannot accomplish your goal through the use of an existing program. Instead of writing a mathematics tutorial program, look into using an existing flexible program such as Mathematica, Maple, or Mathcad. Instead of investing time creating physics software, see if you can accomplish your purpose using Interactive Physics or Working Model. Perhaps many of the goals you intend to fulfill by writing an astronomy program can be satisfied by using the scripting features of Dance of the Planets.

If you are planning a complex project, consider a phased implementation. This will simplify development, as well as allow others to test the application at each phase and provide you with valuable feedback.

Can you count on the help of others who have expertise that might be helpful? Do you have a colleague who has successfully developed an application and can provide guidance and advice about choosing tools and developing the application?

Analysis

You have a clear idea of what your application will do and how. You have ascertained that there is no other program that can solve the problem, so you are gearing up to write your own. Instead of diving into writing code, however, you should continue to develop a plan. The questions that follow may help. Tackling these questions before you write the application will help you develop a better application and avoid problems later.

1. What is the instructional problem you are trying to solve? Perhaps it is to bring more authentic materials within reach of students, or to use the computer to help them visualize a concept that is difficult to grasp.
2. What is the nature of the program? Tutorial? Simulation? Classroom lecture? Tool? Game?
3. List the specific goals that the program is meant to accomplish.
4. In light of the above, what techniques or devices do you plan to use in order to accomplish the goal? Text? Graphics? Animations? Sound? How can you design these so they will be effective? Consider making the program very interactive, to involve the user.
5. What skills will the users of the program already possess? You need to design the program with this question in mind. Consider not only the skills students might have in your discipline. If they are not used to computers, your program is going to have to be very different than if the students use computers on a daily basis.
6. People vary in how they learn. Will this affect the effectiveness of your program?
7. What multimedia will you use? What is your plan for obtaining or developing it?
8. Do you have access to the equipment you will need? Is it powerful enough?
9. Do you have access to the skills you will need?
10. Do you have the funding and time you will need? Programming projects tend to take a lot longer than planned.
11. What equipment is available for students to run the program? Is it capable? It will do little good to rely extensively on software motion video, only to find that it brings the student's machines to their knees, rendering the program unusable. Many people make this mistake, as they develop the program on a fast machine and forget that the machines students have may be much less capable.
12. How will you measure the effectiveness of the program? It might be visually stunning and fun to use, but if it is not accomplishing its educational objectives, what good is it?
13. Have you allowed for a period of testing and program modification?

Learn from others. Do you know of successful programs developed by others? What makes them effective? Have you seen programs that are not effective? Learn from their mistakes! If possible, get ahold of an EDUCOM Software Award–winning program, preferably in your discipline. Programs are judged by experts according to many exacting criteria, including how well they convey the material and their effectiveness as instructional tools.

Design

While analysis looks at the big picture, design concentrates on the details. An important element of design is the program's user interface. No matter how brilliant the program is, it will be less effective if it is difficult to use or inconsistent. You want students to spend their time learning what the program has to teach, not how it works. Here are some suggestions:

1. Always provide status to the students. They should never get lost within your program. It might be useful to provide a means (such as a function key press) for the students to go back to a table of contents or launching point.
2. Make sure the students know what they must do next. They will be frustrated if they get to a point and don't know how to proceed.
3. Make it clear to the students how to provide input. If they are to type text, can they correct it? When they are done typing, do they press the Enter key? How do they get to the next screen?
4. Where appropriate, provide feedback.
5. Consider making help available at the press of a key. At least provide a screen listing the options students can take, in case they forget. You might want to provide means for both forward and backward navigation.
6. In most cases students should be able to go back a screen or two, to review something they might have gone by too fast.
7. The program should have a consistent look and feel from screen to screen. This also applies to use of colors and fonts. Be consistent about how you highlight important information.
8. Do not put too much material on a single screen. Use less than 1/2 of the available screen space. Simple is better. Novice programmers tend to use many colors and fonts, and the results are always unsatisfactory.
9. Consider text size. Is the program to be projected in a classroom or auditorium? Some professors have prepared presentations that the audience could not read because the text was too small when projected in an auditorium.
10. Never use timed screens (unless you are running a psychology experiment). Always let the student control when to proceed. The time for different students to absorb the material on a screen can vary by a factor of 10, so any time value you choose will be too fast for some students, too slow for others.

The Windows environment can help with maintaining a program's consistency. It provides menus, standardized dialog boxes, and so on, which you can use. It is critical to follow Windows conventions. In this way the knowledge students have acquired with other programs helps them learn your program. For example, if your program has a File menu, it should be the first one (on the left) on the menu bar. A Help menu should be rightmost on the menu bar. Many key combinations (Table 8, page 132) are common among Windows programs, and you should implement them where they make sense.

Testing and Evaluation

It is best to test the program on small groups of students during the development process, even if many features are still not implemented. Student feedback is valuable. After all, isn't the program intended for them?

Clearly program bugs will be found and corrected. However, it is also important to find out how easily students can navigate through the program. Do they get lost? Do they always know what to do next? Have they detected inconsistencies in the way the program does things? How does the program run on a variety of computers?

Equally as important, does the program accomplish its educational objectives? Perhaps you can run controlled experiments to determine the difference in learning with and without the use of the program. Were your assumptions valid, or are students less prepared than you expected? Is a particular group of students experiencing significant difficulty? Why? Do students prefer to learn using the program or with traditional methods?

It might be valuable to send the program to a colleague at another campus to use with his or her students. They will probably use the program in ways that your own students will not, thus providing you with additional feedback, suggestions, and program validation.

Implementation

At this point you are satisfied with the program and plan to implement it as a component of a course. It might also be used by your colleagues, and perhaps you are selling or licensing it to other institutions. Because you have been intimately involved with the program from its inception, you might not be aware of assumptions you have internalized.

Is the program documented? What hardware and software requirements does it have? How does it get installed? Is training required? If so, who can provide it? Or, are there training materials available?

What about support? Who can people call with questions? How do they provide feedback? Is the program keyed to a textbook? Have you provided suggestions for its use?

Multimedia Materials

You have a choice when it comes to the multimedia materials you will use in your program. You can use existing materials, or you can create your own. You will be able to save a considerable amount of effort if you can find existing materials to use.

TABLE 8   *Windows Keystroke Conventions*

Action	Keystroke Combination
Quit Program	Alt+F4
Open File	Ctrl+O
Save File	Ctrl+S
Help	F1
Top of File	Ctrl+Home
Bottom of File	Ctrl+End
Extend Selection	Shift
Cancel	Esc

There is a wealth of printed material which, with permission, may be scanned and incorporated into your computer program. Colleges and universities have large collections of videotapes from which new materials may be derived. You or your colleagues might have collections of slides that can be digitized. Another source of material is videodiscs. Many can be "re-purposed," in other words, used for new purposes. For example, a videodisc of a Shakespeare play may be used for teaching English.

Creating your own materials is always a possibility. However, before you commit yourself to following this path, make sure you have the proper equipment and skills. It will be costly if your videodisc is unusable because of the poor quality of the camcorder you used to record the tape, or because of lighting, color, or other problems.

Do not let the difficulties scare you off, though. It does not have to be complicated. For example, a North Carolina State University professor used a portable cassette recorder to tape the calls of different species of owls. The results were quite good, because the subject matter did not require stereo, CD-quality sound. In other words, match the equipment to the "bandwidth" (i.e., quality) of the material you need.

It definitely pays to consult with colleagues who have gone through the process, as they can steer you clear of pitfalls and give you advice too detailed to be included in this chapter. You might also consider attending a couple of workshops, which are offered by a number of institutions. The workshops are especially useful with respect to video, as there is more to consider than with audio, such as camera angles, color balance, lighting, and so on. The advice to attend a workshop also applies to learning about authoring programs and other tools. The Institute for Academic Technology, at Research Triangle Park, North Carolina, for example, offers frequent workshops on various aspects of Multimedia ToolBook program development.

Optimizing Multimedia

A mistake many people make is to try to carry over the techniques from one medium to another. Early applications of computers for instruction tried to imitate the printed page. Students were confronted with page after page of text, interspersed here and there by an occasional question. This technology was so poorly received that the term applied to it, "computer aided instruction" (CAI), acquired negative connotations; new techniques and new terminology had to be sought out. Try to exploit the uniqueness of the medium you are using. Use it where it is most effective.

AUDIO

Recall the significant storage requirements for audio. Use it sparingly. Here are some pointers:

- If you need significant amounts of audio, consider packaging your application on a CD-ROM disc. Too many diskettes will make your application unwieldy.
- Use the lowest recording sample rate you can get away with. Most applications have no need for stereo.
- Consider the delivery environment for your application. A classroom full of computers playing audio will be distracting. Are headphones available?
- Whenever possible use text instead of audio. Spoken instructions will probably be less effective than the same instructions on-screen, where they can

stay until the student has understood them. There are cases, though, where the message is contained as much in the dynamics of the voice as in the contents, such as in the speeches of Dr. Martin Luther King. In cases like this, consider incorporating only the highlights.

VIDEODISC

Here are some points to keep in mind when developing a videodisc.

- Although there is a lot of room on a videodisc, use several short sequences instead of a few long ones. The more the student is actively involved, the better.
- After you get back your videodisc (from the mastering process), consider cutting even more of the material to tighten up your sequences. Just because the frames are on the videodisc does not mean they all have to be used.
- When recording, remember that your video may be shown in a small window on a small screen. Zoom in on the elements of interest. Also pay attention to composition. Elements in the video that do not pertain to the lesson are distracting. Before you start shooting the video, create a storyboard of your program; you should know where the video sequences will fit in and exactly what to shoot.
- Budget at least twice as much as you would expect for your first videodisc. The first one will probably have a number of problems due to your inexperience. Consider it a learning experience.
- After each video sequence is shown to the student, consider adding questions in your program to make sure the student has absorbed the content. Always provide a way for the student to repeat the video. It is easy to do in the program. Remember, the videodisc is different from a videotape: random access is not a problem.
- You might want to provide a way for the student to freeze the frame or play the sequence in slow motion. It might help them visualize the information more easily.
- Videodiscs provide the capability of playing the material at various speeds, and can even play in reverse. Consider ways to exploit this capability. For example, Dr. Stanley Smith's Interactive Chemistry lessons vary the speed of the video in order to match the rate of a chemical reaction to the particular concentration of ingredients the student has chosen.
- Do not use video when a still image will do. Why show 10 seconds of a shaky video of a painting when a still image would be better?

DIGITAL VIDEO

Digital video is very costly in terms of running and storage requirements. Always keep this in mind.

- If you are recording your digital video from an existing videotape, look for every opportunity to cut out information best conveyed by other means. For example, the tape might contain some footage of a professor explaining an experiment. After the explanation, the professor runs the experiment, comments on the results, and runs it again to emphasize some point. Now you are converting the tape to digital video. Discard the explanation and the comments. Instead, put them into text displays in your program. Do not record both occurrences of the experiment. Instead, show the same occurrence twice, or give the student the option to view it as many times as desired. Again, do not think of digital video as a videotape. Think of it as a new medium, with new possibilities and different restrictions.

- Current digital video, especially software motion video, is limited to small video windows. It will rarely achieve full motion, and resolution is poor. Do not rely on it to show much detail.
- Remember that software motion video compensates for a lack of processing power in the player computer by skipping frames. Make sure this will not impair the delivery of the concept you are trying to convey.
- If you do not need sound, record the movies without audio. They will be smaller, and easier to play back.
- In the same manner, try the movie in 256 colors. If it is acceptable, save it in this form, as it will save space and be easier to play back. This does not apply to hardware motion video, which is always stored in 24-bit color format.
- In the unlikely case that you are blessed with lots of memory, you can speed the playback of video for Windows by configuring some of the memory on the playback computer as a RAM disk. Copy the movie (under program control) into the RAM disk before it is needed, and play it from there. Playback performance will be enhanced.

The Future of Multimedia

Trying to predict the future of computer technology is a frustrating undertaking. The technology has developed in unexpected directions, and at a pace that was rarely envisioned. One prediction that can be safely made, however, is that multimedia is here to stay. It is not a fad. Multimedia helps computers communicate with their human users in a more natural way, and thus is useful in many types of applications.

It is difficult to foresee new, revolutionary multimedia technologies. Today's technologies already address the most important human senses: sight (text, graphics, animation, and video), hearing (audio), and touch (e.g., flight simulators). However, the technology is still primitive. Expect great strides in the capabilities of the technology: faster processors, better-resolution graphics, full-screen motion video, fast multimedia networks, and higher-capacity CD-ROMs.

Another technology on the brink of becoming very useful is voice recognition. It is not so much a matter of making it possible to issue menu commands by voice instead of by using the mouse, but a way that people can more easily enter information into the computer. Think of dictating your term paper into the computer instead of having to type it. Think of doctors being able to dictate reports and have them written up, or journalists dictating news stories instead of writing them. In many ways this technology can make significant contributions toward reducing the tedious labor of many.

What makes using a multimedia program fundamentally different from, say, an afternoon at the movies, is interactivity. The advent of personal computers has enabled authors to create highly interactive programs. A further amount of interactivity can be gained by actually immersing the user in the multimedia experience. Virtual reality techniques might be a help in bringing this forth. A variant of this technology is employed in flight simulators used by airlines and the military in pilot training. Surgeons can already practice certain surgical procedures using a powerful computer, virtual reality goggles, and surgical instruments connected to the computer that provide tactile feedback to the surgeon's actions.

Additional research into how we learn will be employed to make multimedia programs more effective in delivering education. Object technology and powerful authoring tools will make programs easier to develop. In less than five years, writing Windows programs has been simplified by several orders of magnitude. New technologies becoming available include visual programming, as exemplified by products such as Visual Smalltalk and Visual Age. These tools let you put together programs by dragging objects onto the working surface, as do ToolBook or Visual Basic, but they also let authors specify interactions between objects by visually connecting the objects together. High-level multimedia authoring tools will make it possible for professors, who heretofore have been unable to invest the time it takes to learn computer programming, to share their knowledge and experience effectively with students through the use of multimedia programs.

In Closing

We have covered the fundamentals of multimedia, including the technology that makes it possible, through the hardware that implements it, and the software that brings it to life.

Multimedia holds tremendous potential to drastically enhance educational effectiveness and the learning experience. Let's close with a quote from multimedia pioneer Dr. William Graves, Associate Provost for Information Technology, University of North Carolina at Chapel Hill, and Director of the Institute for Academic Technology:

> From the student's perspective, the interactive multimedia computer provides a multi-sensory learning environment, an opportunity to learn by commanding a resource that engages all of the senses—to learn by doing. From the scholar's perspective, the multimedia computer provides an opportunity to create, organize, store, retrieve, share and analyze multi-sensory information—sound, graphic and photographic images, motion video, animations, scientific visualizations.

Good luck!

BIBLIOGRAPHY

Alkon, Daniel L. *Memory's Voice: Deciphering The Mind-Brain Code.* New York: HarperCollins, 1992.

Brothers, Hardin. "CD-ROM: Music, Megabytes, Multimedia." *PC Sources,* January 1992.

Brown, C. Wayne, and Barry J. Shepherd. *Graphics File Formats: Reference and Guide.* Greenwich, Conn.: Manning Publications, 1995.

Cartwright, G. Philip. "Teaching with Dynamic Technologies." *Change,* November/December 1993.

Cash, Joan C., ed. *The Power of Multimedia: A Guide to Interactive Technology in Education and Business.* Washington, D.C.: Interactive Video Industry Association, 1990.

Floyd, Steve. *The IBM Multimedia Handbook.* New York: Brady Publishing, 1991.

Gayeski, Diane M., ed. "Getting Started in Multimedia: Avoiding Common Pitfalls." *Multimedia for Learning: Development, Applications, Evaluation.* Englewood Cliffs, N.J.: Educational Technology Publications, 1992.

Hill, Alice, and Thomas Mace. "Optical Explosion." *PC Sources,* October 1991.

IBM Publication GG24-3795-00. "Digital Audio Fundamentals, the M-Audio Adapter, and the Audio Device Driver." Armonk, N.Y.: IBM, 1992.

IBM Publication GG24-3947-00. "Multimedia in a Network Environment." Armonk, N.Y.: IBM, 1993.

Jensen, Robert E., and Petrea Sandlin. *Why Do It? Advantages and Dangers of New Ways of Computer-Aided Teaching/Instruction.* San Antonio, TX: Department of Business Administration, Trinity University, 1991.

Jones, Loretta L., and Stanley G. Smith. "Can Multimedia Instruction Meet Our Expectations?" *Educom Review,* January/February 1992.

Marchionini, Gay. "Hypermedia and Learning: Freedom and Chaos." *Education Technology,* November 1988.

McQuillan, John. "An Introduction to Multimedia Networking." *Business Communications Review,* November 1991.

Smith, Stanley G., and Loretta L. Jones. "The Acid Test: Five Years of Multimedia Chemistry." *T.H.E. Journal IBM Multimedia Supplement,* September 1991.

Yang, Yong-Chil. *The Effects of Media on Motivation and Content Recall.* Farmingdale, N.Y.: Baywood, 1992.

GLOSSARY

Access time The amount of time it takes a disk or CD-ROM drive to start transferring information from the disk or CD-ROM once the location of the information is furnished to the drive.

Active matrix A liquid crystal display technology used in high-end laptop computers. It provides excellent display color.

ADC (Analog-to-Digital Converter) An electronic circuit that converts an analog signal into a digital signal using a process called *sampling*.

ADPCM (Adaptive Differential Pulse Code Modulation) A standard for representing and compressing an audio signal.

Algorithm A step-by-step problem-solving procedure.

Aliasing The effect that results when there are not enough samples to represent the information. It appears, for example, as jagged lines on lower-resolution computer screens and as counter-rotating wheels in film and television.

Amplitude The maximum absolute value reached by the waveform. A measure of strength, volume, or loudness of a signal.

Analog Information represented by a continuous and smoothly varying signal.

ANSI (American National Standards Institute) ANSI Character Set: a standard used by Windows to represent characters, symbols, and control codes (such as a line feed). ANSI uses an 8-bit byte to represent each character, symbol, or control code. The ANSI codes are identical to the ASCII codes for values up to 127, which includes all of the lower- and upper-case letters, digits, and punctuation marks. The ANSI character set includes many foreign and accented characters not present in the ASCII character set.

Archie A program that runs on the Internet; it searches for a file and returns it to the requester. Named after the president of McGill University.

ASCII (American Standard Code for Information Interchange) A standard, widely used in personal computers, to represent characters, symbols, and control codes (such as line feed). ASCII uses an 8-bit byte to represent each character, symbol, or code.

Asymmetric compression A compression process where the time used to compress the information is longer than the time used to decompress the information.

ATM (Asynchronous Transfer Mode) A networking standard characterized by high-speed transmission of small data blocks.

Audio card A hardware card (or board) that you install into a computer and that permits you to record and play digital audio.

Bandwidth The range of frequencies that a device or communications channel can handle.

BASIC (Beginners All-purpose Symbolic Instruction Code) A simplified computer programming language developed at Dartmouth College by Kemeney and Kurtz in the 1960s. Very popular on personal computers.

Bit (Binary Digit) The smallest piece of binary information; it can only have the value of 0 or 1.

Bitmap A form of representing graphical information in which the color value of each one of the picture's pixels (or dots) is stored.

BMP The DOS file extension name for files that contain a bitmap in the Windows or OS/2 format.

Buffer Electronic memory used as temporary storage for data being transferred. The purpose of the buffer is to smooth out the differences in transfer characteristics between the device furnishing the data and the one receiving the data. For example, data might arrive over a communications line or over a network at an instant that the processor is not able to handle it. The data goes into the buffer, because otherwise it would be lost (such data is time dependent). When the processor is able, it can then access the buffer and retrieve the data.

Byte A group of 8 bits treated as a unit.

Card A flat piece of hardware that is plugged into a personal computer slot to perform a specific function, such as to produce sounds, interface to a network, and so on. Also known as a board.

CAV (Constant Angular Velocity) Videodiscs that spin at a constant number of revolutions per minute, meaning that the disc sweeps a constant angle per unit of time.

CDDI (Copper Distributed Data Interface) A high-speed network standard implemented using metal wires.

CD-R (Compact Disc Recordable) A type of compact disc that can be recorded on once by the consumer using a special drive. It does not come recorded from the factory, as do compact discs. Once recorded, a CD-R disc can be played in any drive that accepts standard CD-ROM discs.

CD-ROM (Compact Disc Read Only Memory) A form of compact disc that has computer programs or data recorded on it.

CD-ROM drive A device, attached to a computer, that can read information from a CD-ROM and transfer it to the computer.

CGM (Computer Graphics Metafile) A file format that stores graphics information as a set of drawing commands.

CLV (Constant Linear Velocity) A videodisc where the speed of the track passing by the laser read head is constant. Since the disc track is laid out in a spiral, the r.p.m. of the drive changes, depending on the position of the laser read head; the drive spins at a lower r.p.m. when the laser read head is closer to the outside of the disc.

CMYK (Cyan Magenta Yellow Black) A way of representing the colors of a picture using only four primary colors—cyan, magenta, yellow, and black. Used in color printing.

CODEC (Compressor/Decompressor) An electronic circuit or software program that can compress uncompressed data and decompress compressed data.

Compression Reducing the amount of data it takes to represent information.

DAC (Digital-to-Analog Converter) An electronic circuit that converts a digital sequence into an equivalent analog signal.

Data bus An interface in the computer used to transfer data between devices (such as a disk) and the computer's memory.

Database A collection of data arranged for ease and speed of search and retrieval in one or more computer files. An address book is an example of a database.

Delta modulation A technique for converting an analog signal into a digital signal where the difference in the signal level between samples is represented, as opposed to the current value of the analog signal.

DIB (Device Independent Bitmap) A type of bitmap defined for Windows. Also the DOS file extension for files that contain such a bitmap.

Disk drive A device containing one or more spinning magnetic platters that store information for the computer.

DLL (Dynamic Link Library) A file in Windows or OS/2 that contains program pieces that can be used by other programs. Often parts that are common to a number of programs are placed in DLLs to avoid duplicate storage. When a program needs to run routines that are in a DLL, Windows or OS/2 will automatically load the DLL into the computer's memory and pass control to the requested routine. DLL also refers to the DOS file extension of a dynamic link library file.

DOS (Disk Operating System) A very popular program that controls IBM and IBM-compatible personal computers and enables them to interact with the user and to run other programs.

d.p.i. (dots per inch) A measure of resolution used by printers and display screens. Refers to how many individual dots the device uses to print or display a one-dot-wide line that is one inch long.

DSP (Digital Signal Processor) A specialized computer processor that has circuitry optimized for processing data that represents signal information. Often a DSP is contained in a single chip and is a component of a video digitizing and compression card or of an audio card.

DVI (Digital Video Interactive) A standard for compressing and representing motion video in digital form.

E-Mail (Electronic Mail) Messages sent to other people over computer networks.

Enabling software A program that performs a function on behalf of another program, thus enabling it to perform its task.

Ethernet A very popular technology for connecting computers together across small distances, such as within a building or department. The term "Ethernet" refers to the type of cable used, the hardware that interfaces the cable to the computer, and the protocols used to transfer information over the cable.

Export To move information out of a program to a file, or to send information from one computer to another.

FDDI (Fiber Distributed Data Interface) A high-speed network standard implemented using fiber optic cables.

File A collection of related data or program records stored on media such as disks or diskettes.

File format The way information is stored in a file. The same information can be stored in several different formats, just as a book can be printed in several different languages.

FLC and FLI Two popular formats for files containing computer animations. These formats were developed by Autodesk Corporation.

Font A complete set of type (letters and symbols) of one size and face (appearance).

Frame A single picture in a motion video.

FTP (File Transfer Protocol) A specification for transferring files over the Internet.

Gopher A set of programs that organizes and makes available information over the Internet. The Gopher server provides the information. The Gopher client runs on the user's computer and, at the direction of the user, retrieves information from Gopher servers.

GUI (Graphical User Interface) A program, such as Windows, OS/2, or Macintosh, that enables the user to interact with the computer and its programs via the use of icons, menus, graphical symbols, and a mouse.

Home page The head document in a World Wide Web tree. This document contains links to other documents that the user can access by clicking on highlighted text or pictures. A home page might link to other home pages. The home page is where the user starts the access of related information over the World Wide Web.

Hotword Text that, when clicked with the mouse pointer, causes linked information to be displayed. A hotword can be as small as a single character or as large or larger than a paragraph. By definition, the text that constitutes the hotword is contiguous. Programs highlight hotwords by displaying them in a different color than the rest of the text.

Huffman encoding A form of lossless signal compression in which the most frequently occurring information is stored with as few bits as possible. Named for D. A. Huffman, the inventor of the method.

Hypermedia An organizational structure for presenting

information, where text, graphics, animations, sound, and video are linked together and can be accessed by the user in a nonsequential manner.

Hypertext   Hypermedia, but with text only. Clicking on highlighted text with the mouse cursor causes linked information to be presented.

Hz (hertz)   A measure of the frequency of a signal, in cycles per second. Named in honor of Heinrich Hertz, a German physicist.

Import   To move information into a program from a file, or to bring information into one computer from another.

Interframe compression   Compression of motion video where only the information that changes from frame to frame is saved.

Internet   A worldwide network of computers, composed of thousands of smaller networks around the world, connected together.

Intraframe compression   Compression of the information in a single motion video frame, without reference to, or dependency on, any other frames.

Isochronous   Transferring data over a network in such a manner that sequential portions of the data arrive at the destination at regular intervals.

JPEG (Joint Photographic Experts Group)   An international standard for the compression and encoding of continuous-tone still images (e.g., a photograph). The formal title of the standard is "Digital Compression and Coding of Continuous-Tone Still Images." It is also known as ISO 10918.

KHz (Kilohertz)   One thousand hertz, or one thousand cycles per second.

LAN (Local Area Network)   A means of connecting multiple computers together over short distances (e.g., within a building, a department, or a small campus) that provides high-speed transfer of information among them. Examples of local area networks include Ethernet and token ring.

LCD (Liquid Crystal Display)   A display technology that uses the polarization (orientation) of liquid crystals to either let light through or block it. Liquid crystal displays are popular in watches, calculators, and laptop computers. They are also used in panels that are placed on top of overhead projectors used to display the contents of a computer screen to a larger audience.

Lossless compression   Refers to compression algorithms where data that is compressed can be restored through decompression to its original value. Huffman encoding is an example of lossless compression.

Lossy compression   Refers to compression algorithms where data that is compressed can be restored through decompression to something close to its original value. In other words, some information is lost in the compression/decompression process. Lossy compression algorithms generally result in higher compression (it takes less room to store the compressed data). Users must determine how much loss of information is acceptable. MPEG is an example of lossy compression.

Mainframe   A large, powerful computer, often serving several connected terminals.

MCI (Media Control Interface)   A standard that defines how programs in Windows and OS/2 control multimedia devices.

Megabyte   A million bytes.

Memory   Storage circuitry within a computer. Refers to electronic storage connected to the computer's processor. Programs and data are loaded into memory from a disk before running, because memory is better matched to the speed of the processor than disks are. However, memory is more expensive than the equivalent amount of disk storage, and the information in memory is lost once the power is turned off. This is why, for example, a document in a word processor must be saved to a file before the computer is turned off. Saving the document copies its contents from memory to a disk.

MIDI (Musical Instrument Digital Interface)   A communications standard for representing time-based data for the generation of digital music. Also, the name of the interface that connects the computer to MIDI-capable musical instruments such as synthesizers.

MMM (MacroMind Movie)   A file format used for storing animations created with Macromedia's (formerly MacroMind) Director program. Also the DOS extension for these files.

Mosaic   A software program that is used to retrieve information from the World Wide Web over the Internet. It was developed by the National Center for Supercomputer Applications at the University of Illinois.

MPEG (Moving Picture Experts Group)   An international standard for compression and encoding of information, which includes digital video, digital audio, and timing data. The formal title of this specification is "Coded Representation of Picture, Audio and Multimedia/ Hypermedia Information." It is also known as ISO 11172.

MSCDEX (Microsoft CD-ROM Extensions)   An extension to the DOS operating system, developed by Microsoft, which allows DOS to handle files larger than its 32-megabyte limitation.

Multimedia   The use of text, graphics, animations, sound, and motion video interactively in a computer program.

Multiplexer   An electronic device that places several signals on the same communications channel (wire, fiber, or wireless). It does this either by alternating (in time) the signals sent over the channel (time division multiplexing) or by shifting the frequency of the signals so that each one occupies a different band on the channel (frequency division multiplexing). A demultiplexer on the other end of the communications channel inverts the process, thus reconstructing the original signals. The signals on a cable TV system are an example of frequency division multiplexing. All of the channels are available on the cable at the same time. The user's TV tuner, or cable converter box, serves as the demultiplexer by selecting an individual channel. The TV tuner or cable converter box is not a true demultiplexer, as it only provides one output channel at a time.

Multisession   A CD-ROM drive capable of reading a CD-ROM disc that was recorded in multiple sessions (e.g., a Kodak Photo CD disc), and thus includes multiple tables of contents.

Multispin A CD-ROM drive that can spin the CD-ROM disc faster than what is specified for playing compact disc music. The higher spin rates mean that the CD-ROM tracks go by the laser read head faster, so the transfer rate of the device is faster. Currently available are drives that can spin at 2, 3, 4, or 6 times the normal compact disc spin rate.

NCSA (National Center for Supercomputing Applications) A unit of the University of Illinois that provides high-performance computing resources for universities and corporations. One of NCSA's missions is to aid the scientific research community by producing noncommercial software. The NCSA's work on networking (originally used to connect the supercomputer to remote users) resulted in the development of the Mosaic program for the Internet.

Network A collection of computers that are electronically connected (via wire, fiber optics, telephone, satellite, and so on) for the purposes of communicating. The term "network" generally refers to the elements (cable, cards, software) that are directly involved in transferring information among the connected computers.

NTSC (National Television Standards Committee) The NTSC standard defines the format of signals used for broadcast television in the United States, Canada, and Japan.

Nyquist theorem This states that an analog signal can be sampled with no loss of information as long as the sample rate is at least twice as high as the highest-frequency component in the analog signal. Developed by Harry Nyquist in 1933.

OCR (Optical Character Recognition) A technology that can process a scanned image, recognize text present in the image, and export the text to a text program such as a word processor.

PAL (Phase Alternation Line) The standard for television broadcast signals used by many European countries.

Palette A set of colors or shades (of gray) available to a program for creating its screens. A subset of all of the colors or shades the display screen can display.

Palette shift The condition resulting from the program's need to use more colors for creating a display screen than are available in its palette. Some of the colors displayed will be wrong.

Pantone Matching System (PMS) A system of specifying colors developed by Pantone, Inc. Each Pantone color is assigned a number, and Pantone publishes printed color sets that show each Pantone color along with its number. Graphics programs that implement the Pantone Matching System, such as CorelDRAW! and Fractal Design Painter, allow you to choose colors by specifying the Pantone color number. Print shops can use the Pantone numbers contained in your graphics files to print materials in colors that will exactly match the Pantone colors specified. It is not wise to rely on the colors displayed by the computer's display screen, as these tend to vary some among displays.

Parity A bit that indicates whether the number of 1 bits in a stream is odd or even. Parity is used to check for data transmission errors. It is computed by the sender as the data is sent, and checked by the receiver when the data is received. If the parities generated by sender and receiver do not match, there has been corruption of the data in the transfer.

Passive matrix A liquid crystal display technology used in watches, calculators, and less expensive laptops. Does not have the speed nor the color richness of active matrix displays.

PCM (Pulse Coded Modulation) A digital audio file format obtained by sampling the analog signal and recording the value of the signal at each sample interval.

PCX The DOS file extension name for files that contain a bitmap in a format defined for the PC Paintbrush program. This format was developed by ZSoft Corporation and is a popular bitmap format in DOS and Windows programs.

Period The time interval between the start of two consecutive cycles in a waveform. Defined as 1 divided by the frequency of the signal.

Peripheral device A device inside a computer or attached to the computer that is not a part of the computer's processor or memory. Examples of peripheral devices include disk drives, tape drives, and printers.

Photo CD A CD-R technology and disc format developed by Kodak for the storage of photographs on compact discs.

Pixel (picture element) The smallest addressable element (dot) in a display screen. Each pixel on a display screen can be set to a color independent of all of the other pixels. Computer display technologies are often specified in terms of pixel (or screen) resolution, in other words, how many pixels a display screen shows. A screen with a resolution of 640 × 480 can display 640 pixels per line horizontally and 480 lines from top to bottom. A pixel is not directly related to the size of the dot on the screen. Two display screens, both 640 × 480, will have different pixel sizes if the size of the screen is different.

PLV (Production-Level Video) A video compression format using DVI technology. PLV is an asymmetric compression scheme. The video to be compressed is digitized from a videotape and processed by a computer for as long as it takes to achieve optimal compression.

Processor The electronic unit inside a computer that executes instructions and manipulates (processes) data. Computer models from different manufacturers might have the same processor. Examples of processors include the Intel 486 and the Motorola 68000.

Protocol A set of rules governing how communication and data transfer takes place over a network or interface. A protocol is analogous to a spoken language's grammar. In English, for example, grammar specifies the words we can use and how to put these words together to express thoughts, which includes rules about forming plurals, conjugating verbs, and so on. A network protocol, for example, specifies the form and the size (how many bytes can be transferred in one sitting) of data on the network, how to send and receive data, how to specify who the recipient of the data is, how to handle transmission errors, and so on.

Read only   Media, such as CD-ROM discs or videodiscs, that are recorded at the factory and cannot be erased or re-recorded by the user.

RGB (Red Green Blue)   A method in which colors are created by combining only red, green, and blue primary colors. RGB color is used in computer screen displays and for video (television, camcorders); colors are displayed by these devices by lighting red, green, and blue dots on the screen.

RLE (Run-Length Encoding)   A form of lossless signal compression in which patterns (runs) of data values are recognized. The data is compressed by replacing the repetitive data sequence with a pattern and a count. Run-length encoding is used to compress fax data, so there is less data to send over the phone line than would be the case if each dot of the original were transmitted. Several popular personal computer file formats employ run-length encoding to represent the data more efficiently. Examples include PCX and TIFF files.

RTV (Real-Time Video)   A format for video compression supported by the DVI technology. RTV video is compressed as fast as the video signal is received. Thus it is a symmetric compressor.

S-VHS   A derivative of the VHS video recording standard used in consumer video cassette recorders. It provides higher-quality video than that provided by standard VHS. S-VHS can also refer to the signal provided by an S-VHS–capable VCR or camera.

Sampling   The process of converting an analog signal to digital form by periodically noting and storing (sampling) the value of the analog signal.

Scalable font   A typeface that can be scaled by the computer to any reasonable point size. Examples of this technology include Adobe Type 1 and TrueType.

SCSI (Small Computer Systems Interface)   An interface standard that provides for the attachment of some peripheral devices, such as disk drives and CD-ROM drives, to the computer.

Serial communications   Communication that takes place using the serial port of the computer, generally through a modem to a phone line. Only 1 bit of information is transferred at a time.

Serial port   An interface connection present on most personal computers that permits communication.

Server   A computer connected to a network that provides resources to other computers on the network. A file server stores files on its own disks that can be accessed by other computers on the network. A print server accepts print jobs from other computers on the network and sends them to printers connected to the server.

Synthesizer   An electronic device that takes input in the form of MIDI signals, or signals from a keyboard, and generates audio signals in response. The signals produced by the synthesizer can be routed to an amplifier and speakers to produce sound.

Symmetric compression   A compression process where the time used to compress the information is the same as the time used to decompress the information.

TCP/IP (Transfer Control Protocol/Internet Protocol)   A communications protocol used for the Internet and also used in many campus networks.

TelNet   A protocol used over the Internet that allows a user to log on to a remote computer connected to the Internet.

TIF (Tag Image File)   The DOS file extension for files in TIFF format.

TIFF (Tag Image File Format)   A file format for files containing bitmaps. This format is widely used in desktop publishing applications. It was developed by Aldus Corporation in cooperation with several other companies. Each data item in a TIFF file has a unique tag associated with it. Data items, in addition to the bits specifying the image itself, include image width, image length, bits per sample, color map, program that created the file, and many others. If a program is not interested in a specific tag in the file, it can skip it. Information from a TIFF file can be retrieved randomly, so it is not necessary, for example, to read the whole file in order to change the values of a small portion of the image.

Token ring   A popular technology for connecting computers together across small distances, such as within a building or department. This type of network continuously passes a token (a specific sequence of bytes) around the network. When a computer on the network needs to transfer information, it intercepts the token and places its data on the network. Data collisions (multiple computers trying to transmit at the same time) are avoided because only the computer that currently owns the token can place data on the network.

Transfer rate   The amount of data that a network or interface can transfer in a unit of time.

TWAIN (Technology Without An Interesting Name)   A protocol for communication between scanners and computers.

Typeface   Type of the same design. For example, Times Roman is a typeface. Twelve-point Times Roman normal is a font, as is 12-point Times Roman bold or 14-point Times Roman normal.

UNIX   An operating system (the program that controls the computer) developed by Bell Labs and now owned by Novell. UNIX ran on small computers hooked up to a number of terminals that served a small group such as a department. Currently UNIX runs on most workstations (single-user computers that employ processors more powerful than those used by personal computers).

Veronica (Very Easy Rodent-Oriented Netwide Index of Computerized Archives)   A program that runs on the Internet, which serves as a directory for Gopher servers.

VGA (Video Graphics Array)   The name of a display standard for personal computers. VGA is specified as a display that can show 16 colors at a resolution of 640 × 480. A newer standard, Super VGA or SVGA, is defined to display 256 colors at a resolution of 800 × 600. The minimum acceptable color depth for multimedia is 256 colors.

WAV The DOS extension of files that contain digital audio in PCM format. WAV is short for waveform.

WMF (Windows Metafile) The DOS extension of files that contain graphics in the Windows metafile form, a form that specifies graphics by encoding the set of graphic commands necessary to recreate the graphic.

WWW (World Wide Web) A wide-area hypermedia information-retrieval initiative aiming to give universal access to a large universe of documents. The World Wide Web consists of interlinked information (content), which is made available over the Internet through Web servers. It is accessed over the Internet using Web client programs such as Mosaic.

XA (Extended Architecture) A feature of newer CD-ROM drives that supports interleaving of files, in other words, placing data from multiple files alternately on the same track of the disc, to allow multiple files to be read at once.

INDEX

INTEGRATED MEDIA GROUP

An Imprint of Wadsworth Publishing Company

I(T)P™ An International Thomson Publishing Company

Belmont • Albany • Bonn • Boston • Cincinnati • Detroit • London • Madrid • Melbourne •
Mexico City • New York • Paris • San Francisco • Singapore • Tokyo • Toronto • Washington

Technology Publisher: Kathy Shields
Assistant Editor: Tamara Huggins
Production Services Coordinator: Gary Mcdonald
Production: Greg Hubit Bookworks
Print Buyer: Karen Hunt
Designer: Stuart Paterson/Image House, Inc.
Copy Editor: Michele Jones
Cover Designer: Ellen Pettengell
Compositor: Monotype Composition Company, Inc.
Printer: Courier Companies/Kendallville
Cover Printer: Phoenix Color Corp.

Printed in the United States of America
1 2 3 4 5 6 7 8 9 10—01 00 99 98 97 96 95

For more information, contact Wadsworth Publishing Company:

Wadsworth Publishing Company
10 Davis Drive
Belmont, California 94002, USA

International Thomson Publishing Europe
Berkshire House 168-173
High Holborn
London, WC1V 7AA, England

Thomas Nelson Australia
102 Dodds Street
South Melbourne 3205
Victoria, Australia

Nelson Canada
1120 Birchmount Road
Scarborough, Ontario
Canada M1K 5G4

International Thomson Editores
Campos Eliseos 385, Piso 7
Col. Polanco
11560 México D.F. México

International Thomson Publishing GmbH
Königswinterer Strasse 418
53227 Bonn, Germany

International Thomson Publishing Asia
221 Henderson Road
#05-10 Henderson Building
Singapore 0315

International Thomson Publishing Japan
Hirakawacho Kyowa Building, 3F
2-2-1 Hirakawacho
Chiyoda-ku, Tokyo 102, Japan

Library of Congress Cataloging-in-Publication Data

Pinheiro, Edwin J.
    Introduction to multimedia : featuring Windows applications / Edwin
  J. Pinheiro.
        p.   cm.
      ISBN 0-534-26634-7
      1. Multimedia systems.   2. Windows (Computer programs)   I. Title.
  QA76.575.P56   1996
  006.6—dc20   95-23941

Introduction to
Multimedia

Featuring

WINDOWS

APPLICATIONS

EDWIN J. PINHEIRO

IBM Academic Consulting & Services

INTEGRATED MEDIA GROUP

An Imprint of Wadsworth Publishing Company

I(T)P™ An International Thomson Publishing Company

Belmont • Albany • Bonn • Boston • Cincinnati • Detroit • London • Madrid • Melbourne •
Mexico City • New York • Paris • San Francisco • Singapore • Tokyo • Toronto • Washington

Technology Publisher: Kathy Shields
Assistant Editor: Tamara Huggins
Production Services Coordinator: Gary Mcdonald
Production: Greg Hubit Bookworks
Print Buyer: Karen Hunt
Designer: Stuart Paterson/Image House, Inc.
Copy Editor: Michele Jones
Cover Designer: Ellen Pettengell
Compositor: Monotype Composition Company, Inc.
Printer: Courier Companies/Kendallville
Cover Printer: Phoenix Color Corp.

Printed in the United States of America
1 2 3 4 5 6 7 8 9 10—01 00 99 98 97 96 95

For more information, contact Wadsworth Publishing Company:

Wadsworth Publishing Company
10 Davis Drive
Belmont, California 94002, USA

International Thomson Publishing Europe
Berkshire House 168-173
High Holborn
London, WC1V 7AA, England

Thomas Nelson Australia
102 Dodds Street
South Melbourne 3205
Victoria, Australia

Nelson Canada
1120 Birchmount Road
Scarborough, Ontario
Canada M1K 5G4

International Thomson Editores
Campos Eliseos 385, Piso 7
Col. Polanco
11560 México D.F. México

International Thomson Publishing GmbH
Königswinterer Strasse 418
53227 Bonn, Germany

International Thomson Publishing Asia
221 Henderson Road
#05-10 Henderson Building
Singapore 0315

International Thomson Publishing Japan
Hirakawacho Kyowa Building, 3F
2-2-1 Hirakawacho
Chiyoda-ku, Tokyo 102, Japan

Library of Congress Cataloging-in-Publication Data

Pinheiro, Edwin J.
    Introduction to multimedia : featuring Windows applications / Edwin
J. Pinheiro.
       p.   cm.
    ISBN 0-534-26634-7
    1. Multimedia systems.   2. Windows (Computer programs)   I. Title.
QA76.575.P56   1996
006.6—dc20   95-23941

CONTENTS